The Last Billion Years

A Geological History of the Maritime Provinces of Canada

Atlantic Geoscience Society

Authorship: Atlantic Geoscience Society.

Publisher: Nimbus Publishing Ltd., P.O. Box 9166, Halifax, NS B3K 5M8
Atlantic Geoscience Society

Designer: Neil Meister, Semaphor Design Company, Halifax, NS

Printer:

The Last Billion Years Book Committee

Chair: Graham L. Williams

Project Leader: Robert A. Fensome

Production Coordinator: Jennifer L. Bates

Fund-Raising Coordinator: David E. Brown

Members: Sandra M. Barr, John H. Calder, Howard V. Donohoe, Jr., Warren B. Ervine, Laing Ferguson, Leslie R. Fyffe, Martin R. Gibling, Robert G. Grantham, Andrew S. Henry, Michel P. Latremouille, R. Andrew MacRae, Henrietta Mann, Randall F. Miller, Ronald K. Pickerill, David J.W. Piper, John W. Shimeld, Ralph R. Stea, John A. Wade, Peter I. Wallace, Reginald A. Wilson

ISBN: 1-55109-351-0

Atlantic Geoscience Society Special Publication No.15

Contributors

Editors: Robert A. Fensome, Graham L. Williams.

Technical Editor: Clare Goulet.

Editorial Panel: Robert A. Fensome, Erin Frith, Michel P. Latremouille, Graham L. Williams.

Chapter Authors
Sandra M. Barr, John H. Calder, Robert A. Fensome, Leslie R. Fyffe, David J.W. Piper, Ralph R. Stea, John A. Wade, Graham L. Williams, Reginald A. Wilson.

Box Authors
Kenneth D. Adams, Jennifer L. Bates, John H. Calder, D. Barrie Clarke, Warren B. Ervine, Robert A. Fensome, Robert G. Grantham, Randall F. Miller, Antonius G. Pronk, John Shaw, Robert B. Taylor, Graham L. Williams, Reginald A. Wilson.

Text Contributors
Ross R. Boutilier, Warren B. Ervine, Laing Ferguson, Martin R. Gibling, Robert G. Grantham, Daniel J. Kontak, R. Andrew MacRae, Henrietta Mann, Gwen L. Martin, Francine M. McCarthy, Steven R. McCutcheon, Randall F. Miller, Ronald K. Pickerill, Garth A. Prime, Antonius G. Pronk, Allen A. Seaman, John W. Shimeld.

Frontispiece: The dramatic effects of the world's highest tides on Triassic sedimentary rocks in the Bay of Fundy. Paddys Island, NS

Text Advisors
Dominique Bérubé, Robert C. Boehner, Fred Bonner, John R. De Grace, Donald L. Forbes, Christopher D. Jauer, Krista L. McCuish, Peta J. Mudie, J. Brendan Murphy, Paul E. Olsen, Michael A. Parkhill, Clint St. Peter, John W.F. Waldron, Peter I. Wallace, J.B.W. (Hans) Wielens.

Photographs: Sandra M. Barr, Wayne Barrett, Dominique Bérubé, Robert C. Boehner, David E. Brown, John H. Calder, Donald J.J. Carroll, John R. De Grace, Howard V. Donohoe, Jr., Robert A. Fensome, Laing Ferguson, Leslie R. Fyffe, Ronald Garnett, Patricia G. Gensel, Martin R. Gibling, Robert G. Grantham, Andrew S. Henry, Lubomir F. Jansa, John P. Langton, William C. MacMillan, R. Andrew MacRae, Martin E. Marshall, Gwen L. Martin, Krista L. McCuish, Steven R. McCutcheon, Ronald Merrick, Brent V.Miller, Randall F. Miller, Alan V. Morgan, Clarence Nowlan, Gordon N. Oakey, Michael A. Parkhill, Ronald K. Pickerill, Antonius G. Pronk, Robert P. Raeside, Gerald E. Reinson, Arie A. Ruitenberg, Clint St. Peter, Allen A. Seaman, Ian S. Spooner, Ralph R. Stea, Robert B. Taylor, Jacques J. Thibault, Gilbert van Ryckevorsel, Keith Vaughan, Gerard J.M. Versteegh, John W.F. Waldron, John Wm. Webb, Heinz F. Wiele, Alex A. Wilson.

Paintings: Judi O. Pennanen, Stephen F. Greb.

Line Drawings: Christopher Hoyt.

Graphics: Jonathan P. Bujak, John H. Calder, Steven P. Colman-Sadd, Howard V. Donohoe, Jr., Robert A. Fensome, Laing Ferguson, Martin R. Gibling, Robert G. Grantham, Andrew S. Henry, Christopher Hoyt, William C. MacMillan, Ronald F. Macnab, Conall Mac Niocaill, R. Andrew MacRae, Francine M. McCarthy, Gordon N. Oakey, David J.W. Piper, Antonius G. Pronk, Walter R. Roest, Christopher R. Scotese, Susan A. Scott, John Shaw, Ralph R. Stea, Cees R. Van Staal, John A. Wade, Graham L. Williams, Valerie J. Williams, Reginald A. Wilson.

Advisors: Kenneth D. Adams, Sandra M. Barr, Ross R. Boutilier, Tim J. Fedak, Leslie R. Fyffe, Ronald F. Macnab, Paul E. Olsen, Clint St. Peter.

Source Materials: Ted Daeschler and the Academy of Natural Sciences (Philadelphia), Gordon Dickie, Robert A. Fensome, Andrew Flanagan, Fundy Geological Museum, Patricia G. Gensel, Eldon George, Geological Survey of Canada (Atlantic), Peter Hacquebard, Stephen Horne, Eric Leighton, William C. MacMillan, R. Andrew MacRae, David Mossman, New Brunswick Museum, Nova Scotia Museum of Natural History, Gordon N. Oakey, Gerry R. Oram, Alison Lodge and Parks Canada, Redpath Museum of McGill University, Donald Reid, Alan Ruffman.

Reviewers: Warman F.P. Castle, Janet Jauer, John A. Wade, Jeanette M.J. Wielens.

Technical Support: Katherine M. Goodwin, Gary M. Grant, Andrew S. Henry, Nelly Koziel, Terry L. Leonard, William C. MacMillan, Suzanne C. Purdy, Rhonda L. Sutherland.

Table of Contents

Benefactors

Publication of "The Last Billion Years" has been made possible
through the generous contributions of the following:

Sponsors
($5,000–$20,000 or in-kind equivalent)

Canadian Geological Foundation

Geological Survey of Canada (Atlantic)

Natural Resources Canada

New Brunswick Museum

Nova Scotia Department of Natural Resources

PanCanadian Petroleum Limited

Sable Offshore Energy Incorporated

Supporters
($1,000–$4,999)

Atlantic Geoscience Society

Canadian Institute of Mining, New Brunswick Branch

Canadian Society of Petroleum Geologists

Dalhousie University

Geological Society of the Canadian Institute of Mining

Marathon Canada Limited

Noranda Mining and Exploration Inc.,
New Brunswick Branch

Petroleum Society of the Canadian Institute of Mining

Shell Canada Limited

Donors
(up to $999 or in-kind equivalent)

Blair MacKinnon

Cardinal Communications Ltd.

Elmtree Resources

Halifax Section—Petroleum Society of the Canadian Institute of Mining

Nova Gold Resources Inc.

Teachers' Inservice, Nokomis High School, Maine

Introduction

A scene from the ancient Maritimes. In early Carboniferous lakes (about 360 million years ago) there were a variety of fish. In this reconstruction, based on fossils found at Albert Mines, NB, are three types of paleoniscoid fish: the silvery-purple ones (*Rhadinichthys*), in a school swimming toward the left; the larger brownish-grey ones (*Elonichthys*), three of which are seen swimming toward the right; and, at the far right, a single specimen of a deep-bodied paleoniscoid similar to *Eurynotus*. A large, lobe-finned fish (*Latvius*) is also present at left-centre, above the paleoniscoid school.

The Last Billion Years is a history book. It is not a conventional sort of history book, however, but one that highlights the geological history of the Maritime Provinces of Canada. A conventional history would deal with centuries, or perhaps millennia, whereas this book deals with hundreds of millions of years. Indeed, in the Maritime Provinces, we have direct evidence for a geological history stretching back about 1,000 million (or a billion) years. The Earth has changed a lot in the last billion years. Continents have moved across the globe, collided and split apart. Oceans have been formed and destroyed. Climates have wavered between tropical and frigid. And simple life forms have evolved into multi-celled animals and plants, some of which have colonized the land. The story of these changes, as they have affected the Maritime Provinces, is the major plot of this book.

The Maritime Provinces (or simply "the Maritimes") is the collective name given to New Brunswick, Nova Scotia and Prince Edward Island. Generally, we visualize these provinces as bounded by seemingly unchanging coastlines. However, coastlines are ephemeral features in a geological sense. Erosion by waves and tides, the silting up of an estuary or the extension of a sand spit can markedly change a coastline, even within the span of a human life. Over tens of thousands and millions of years, coastlines change beyond recognition, and the boundary between land and sea can "regress" far out onto the continental shelf or "transgress" hundreds of kilometres onto the continent. So the present-day offshore areas—the Scotian Shelf, the Bay of Fundy and the Gulf of St. Lawrence—are an integral part of our story. This is justifiable in a social sense too; Maritimers consider such things as Fundy tides and offshore fishing and oil exploration as very much part of their identity.

Why write a book about the geological history of the Maritimes? First, the region's varied landscapes and geology reflect a fascinating past that is known to specialists, but no comprehensive modern account of this history is available for the non-specialist. Second, geology has played a key role in determining where people live, build roads and dispose of waste, where resources such as water, coal, oil and metals can be found, and where there are fertile soils. Third, the geological history of the Maritimes is particularly varied and interesting, and there are several internationally important geological locations within the region.

We hope that those interested in geology who live in, visit, or want to learn about the Maritimes will benefit from reading this book. We have tried to keep jargon to a minimum, but sometimes it is needed to avoid awkward or repetitive descriptions. However, we explain all technical words where we first use them, and we illustrate concepts with artwork, diagrams and photographs.

When we started writing this book, we wanted to take the reader back one billion years in the first chapter. We soon realized, however, that we needed to set the scene by providing an introduction to geology. Hence, we do this in the first three chapters, using Maritime examples where possible. Chapter 1, **The Dynamic Earth**, explains earth systems such as plate tectonics, the rock cycle and the changing climate. In Chapter 2, **The Fourth Dimension**, we discuss time–how we know the age of rocks, and how we correlate rocks of the same age from one place to another. Chapter 3, **Tales of Trails and Ancient Bodies**, highlights the fascinating story of fossils, how they are preserved and how they relate to present day life. Those with a grounding in geology may wish to skip the first three chapters. **From Rocks to Riches** (Chapter 8), demonstrates the importance of natural resources to the Maritimes and how such commodities as gold, coal, oil, natural gas, gypsum, building stones, and sands and gravels are found. We also introduce some environmental issues.

All other chapters describe the geological history of the Maritimes, beginning with the oldest rocks and fossils in Chapter 4, **Into Deepest Time**. Here we also discuss the earliest known supercontinent, Rodinia, to which a small part of the current Maritimes belonged. In Chapter 5, **The Pieces Come Together**, we show how a patchwork of crustal fragments, including some from the southern hemisphere, came together to form much of the Maritimes. We also show how two oceans, the Iapetus and the Rheic, arose and were later destroyed. Chapter 6, **Basins and Ranges**, highlights a time when tropical seas and coal forests spread into our region, with the Maritimes at the heart of a huge supercontinent known as Pangea. We relate how some of the earliest reptiles were found in our region and how they lived in a world with huge dragonfly- and millipede-like creatures. Chapter 7, **An Ocean is Born**, traces events related to the origin of the Atlantic Ocean. There was no Atlantic Ocean 200 million years ago, and the Maritimes was adjacent to northwest Africa, within the former continent of Gondwana. When the Atlantic opened, a piece of ancient Gondwana was left behind on the North American side. Hence, it could be argued that, in a geological sense, Halifax is an African city. In Chapter 9, **The Ice Age and Beyond**, we discuss the relatively recent past, the last million years, a time when such creatures as mastodons and walruses lived in the Maritimes. As recently as 20,000 years ago, this region was covered by glaciers that shaped our modern landscape; plants and animals living in the Maritimes today all migrated here after the ice retreated. We finish our story by gazing into a crystal ball to try to predict when there will be further ice ages and new mountain ranges in our region.

Within each chapter, we include one or more "boxes" on subjects that deserve a more detailed account than is possible in the main text. These boxes cover a wide range of topics, including the structure of the Earth, famous Maritimes geologists, minerals, fossil fish, granites and coastlines. To close the book, we list resources (e.g. museums, books, pamphlets, maps, videos and web sites) from which readers can glean further details. We hope that this book will whet the reader's appetite for a deeper exploration of the Maritimes and its geological past—our common heritage.

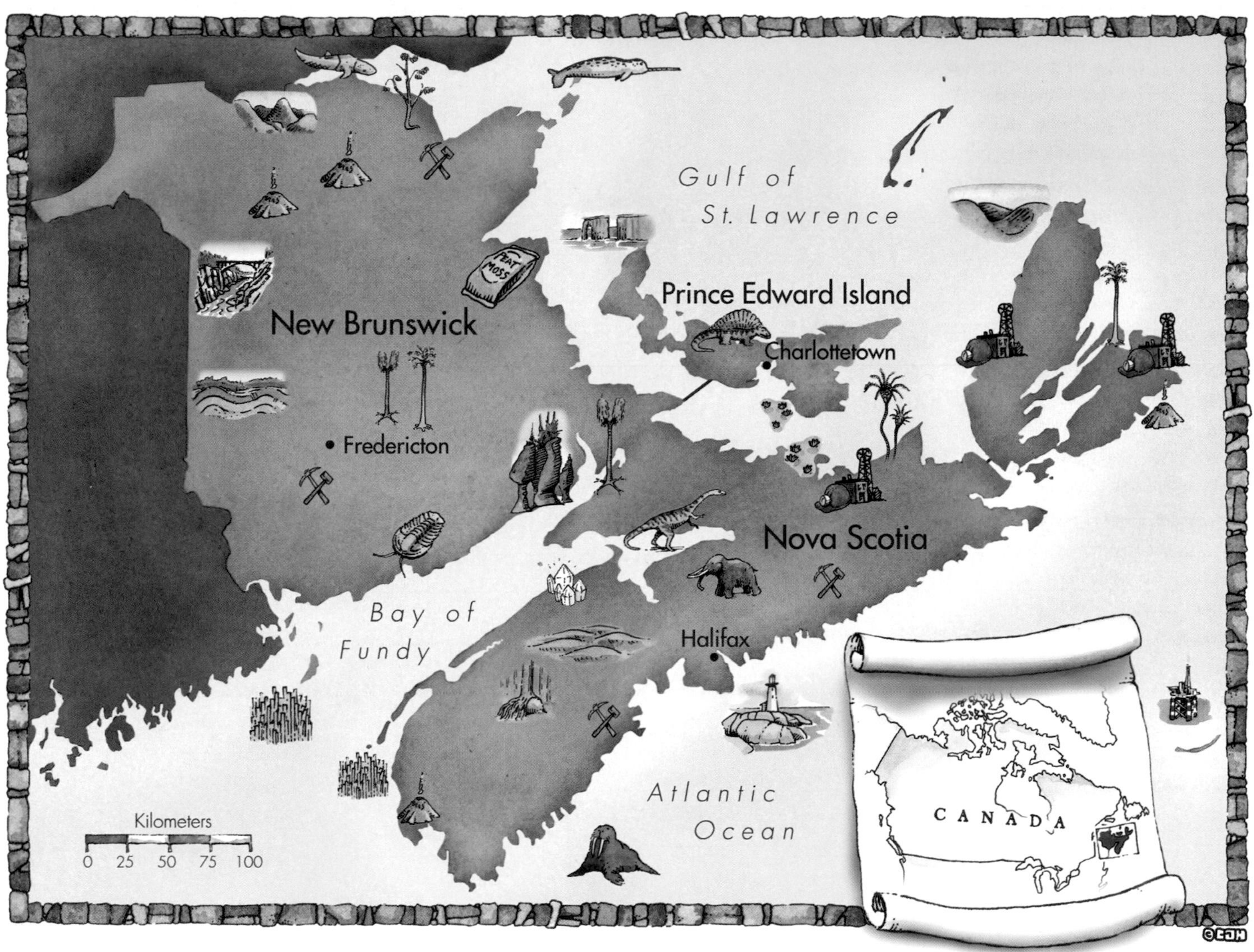

Geological highlights of the Maritime Provinces of Canada include landscapes molded by the recent Ice Age, relics of ancient, long-extinct volcanoes, world-famous mineral and fossil localities, the exploration for and the extraction of natural resources, and scenic attractions.

New Brunswick is abundantly stored with valuable minerals that are capable of supplying the means of strength and greatness.

—Abraham Gesner, in his Fourth Report on the Geological Survey of the Province of New Brunswick, 1842.

I never travelled in any country where my scientific pursuits were better understood or more zealously forwarded than in Nova Scotia.

—Sir Charles Lyell, Travels in North America, 1842.

The red sandstone everywhere supports a fine, friable, loose, loamy soil, which renders Prince Edward Island one of the finest agricultural districts in the lower provinces ....

—Sir John William Dawson, Acadian Geology, Second Edition, 1868.

Erupting volcano—a dramatic example of the dynamic Earth.

CHAPTER 1
The Dynamic Earth

Drifting Continents, Wandering Poles and Spreading Oceans

Early on the morning of 14 November 1963, the fishing vessel *Ísleifur II* was plying North Atlantic waters, about four nautical miles off the Vestmann Islands, south of Iceland. After morning coffee, engineer Ârni Gudmundsson and skipper Gudmar Tómasson noticed a strange, sulphurous odour. Then, mysteriously, the boat started to roll. The cook, Ólafur Vestmann, spotted a dark column of smoke to the southeast and thought that another ship was on fire. However, after confirming that no SOS calls had been received, the captain suspected volcanic activity and contacted the authorities ashore.

The skipper was correct. The crew of the *Ísleifur II* had witnessed the opening scenes in the creation of a new volcanic island: Surtsey. Over the next few hours, the eruption increased in power. By mid-afternoon, the airborne column of volcanic material was visible in Iceland's capital, Reykjavik, 110 kilometres away. Within a week, the new island was 43 metres high and over 600 metres long. Eruptions on and around Surtsey continued until the autumn of 1965, when the island was more than 2 kilometres across.

Volcanic activity in the oceans is common. Islands along the entire length of the Atlantic Ocean—the Azores, St. Paul Rocks near the equator, Napoleon's final island-prison St. Helena, Tristan da Cunha, and Bouvet Island just north of the Antarctic ice pack—are all volcanic in origin. And Iceland itself is the planet's largest volcanic island, made up of dozens of individual volcanoes. In fact, the islands of the mid-Atlantic, from Iceland in the north to Bouvet Island in the south, are linked by one of the earth's longest mountain chains, the Mid-Atlantic Ridge, which lies mostly beneath the ocean. And, except for a thin veneer of sediment, this 16,000-kilometre-long mountain chain is entirely volcanic. Moreover, we now know that most rocks on the ocean floor are volcanic in origin. These oceanic rocks are typically less than 135 million years old. This is in great contrast to the age of many rocks forming the continents, which can be as old as 3.5 billion years.

The Atlantic sea floor becomes older away from the Mid-Atlantic Ridge. The ages are based on the sequence of magnetic polarity reversals.

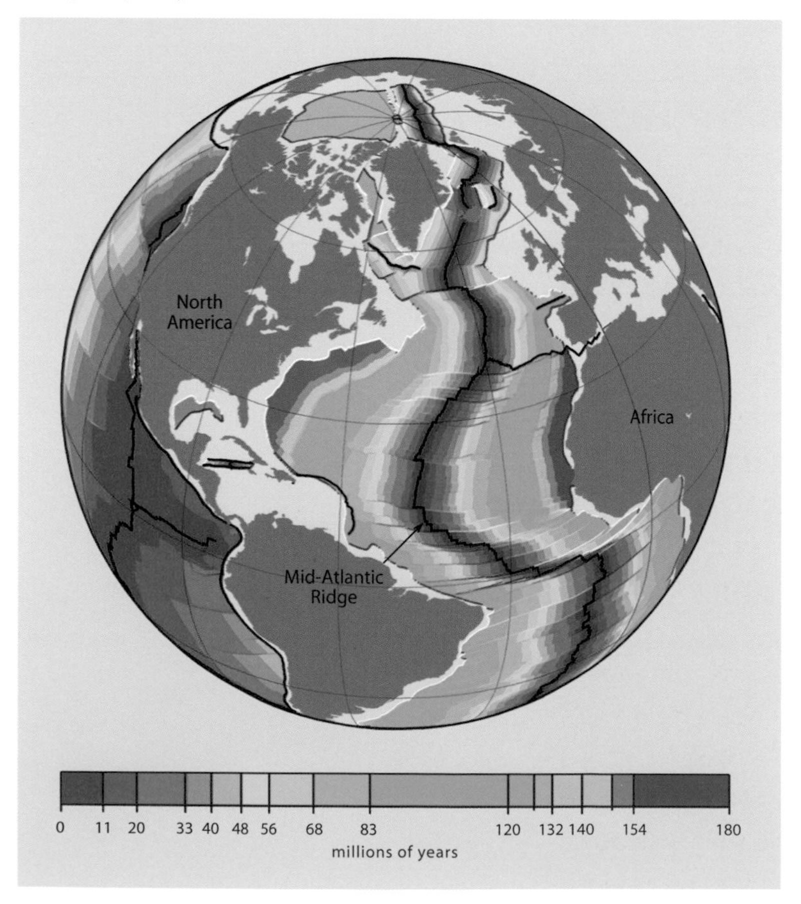

Why is the ocean floor mainly volcanic? And why is it so young? To answer these questions, geologists have developed a theory called plate tectonics. This theory evolved out of the realization that continents and oceans are neither fixed nor permanent. Rather, continents have moved or drifted over the globe, and oceans have opened and closed during the past billion years or more. This is because the Earth's outer layer is divided into a series of rigid plates that constantly move in relation to each other. But how did the theory of plate tectonics develop?

As long ago as 1620, Sir Francis Bacon (1561-1626) recognized that the coastlines of South America and Africa would fit snugly together if the South Atlantic Ocean were closed up. At the beginning of the twentieth century, the German

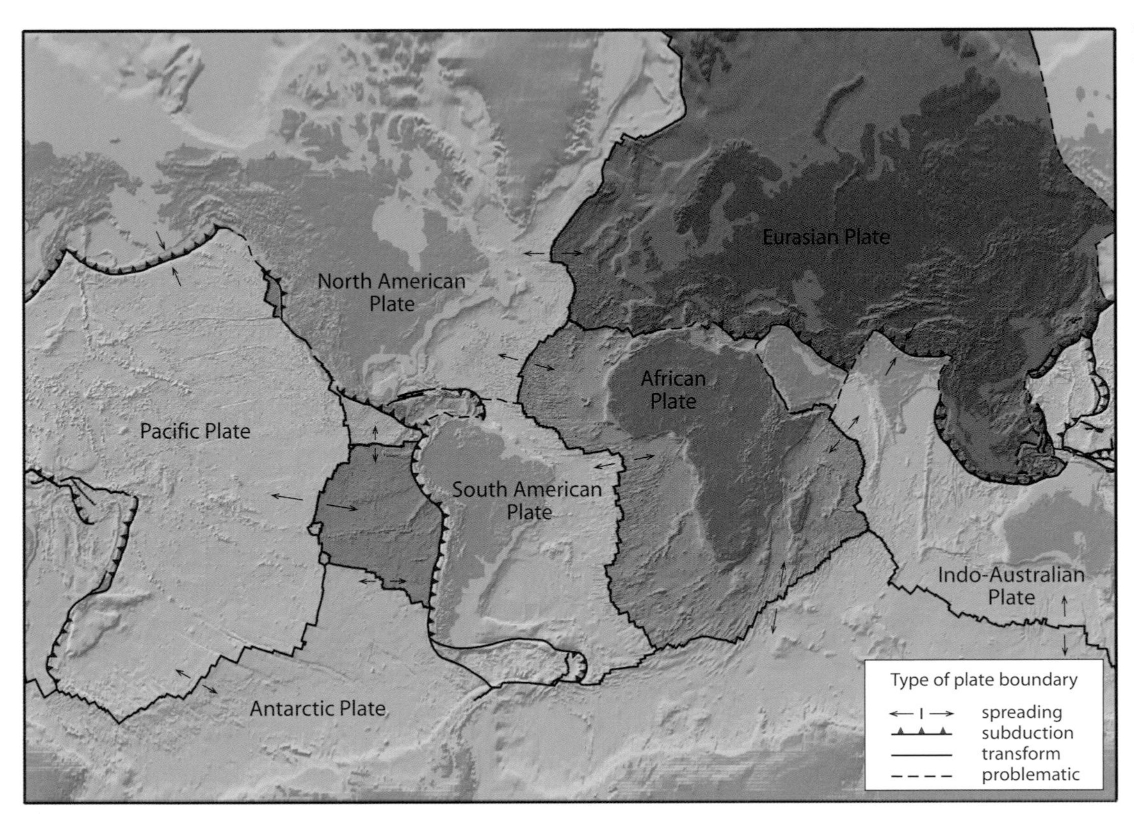

The Earth's present-day tectonic plates.

meteorologist Alfred Wegener (1880-1930) was also impressed by the trans-Atlantic continental fit. He collected other evidence to support the idea that continents had moved over time. A good example is the distribution of the fossil freshwater fish-eating reptile *Mesosaurus*, which lived in Brazil and southern Africa about 250 million years ago, but apparently nowhere else. Wegener compared the geological structures along the coasts of Brazil and southwest Africa: "It is just as if we were to refit the torn pieces of a newspaper by matching their edges." He observed glacial deposits in rocks that were about 300 million years old. These rocky archives of an ancient deep freeze are found only in India, Australia, Africa and South America. They were deposited at the same time that the Maritimes was enduring humid, tropical conditions in which coals were forming. This distribution of glaciers and tropics is impossible to explain if the continents have always been in their present positions.

When World War One broke out in 1914, Wegener was conscripted into the German army and later wounded. While convalescing, he wrote a book on his geological research, published in English in 1924 as *The Origin of Continents and Oceans*. Thus was born the theory of continental drift. However, most geologists of Wegener's time opposed this theory. Because there was no known mechanism to explain the movement of continents, the geological establishment could not accept these ideas from a meteorologist, no matter how compelling the evidence. To explain Wegener's observations, geologists came up with complicated explanations, such as the presence of land bridges between continents. And so the debate continued for several decades.

A major breakthrough was made in the 1950s by geophysicists, scientists who study the structure of the Earth through its physical properties, such as gravity and magnetism. They showed that planet Earth behaves like a giant magnet. The Earth's core contains dense matter made up mainly of nickel and iron. Motion in the outer core, which is probably liquid, generates a magnetic field. When iron-containing minerals cool and crystallize to form rocks, they become magnetized like a compass needle, taking on the existing magnetic alignment at the time they solidify. Measurements of these "paleomagnetic" properties can be used to find the geographic latitude of the rock when it formed, as well as the position of the Earth's magnetic pole at that time.

Such measurements were crucial in proving Wegener correct. The English geophysicist Stanley Runcorn and his colleagues examined the paleomagnetic record from European volcanic rocks. They found that the apparent position of the magnetic North Pole was different for different ages of rock. For example, data from rocks that formed about 500 million years ago indicate that the pole, at that time, was situated in what is today the western Pacific Ocean, whereas data for rocks about 100 million years old suggest that the pole was situated somewhere around northern Siberia. This information was plotted on a map to produce a so-called "polar wandering curve". This curve showed that either the poles had wandered over time or that the continents had drifted (or a combination of both). New evidence soon narrowed the options and, indeed, forced a conclusion.

"Wandering" magnetic poles through time.

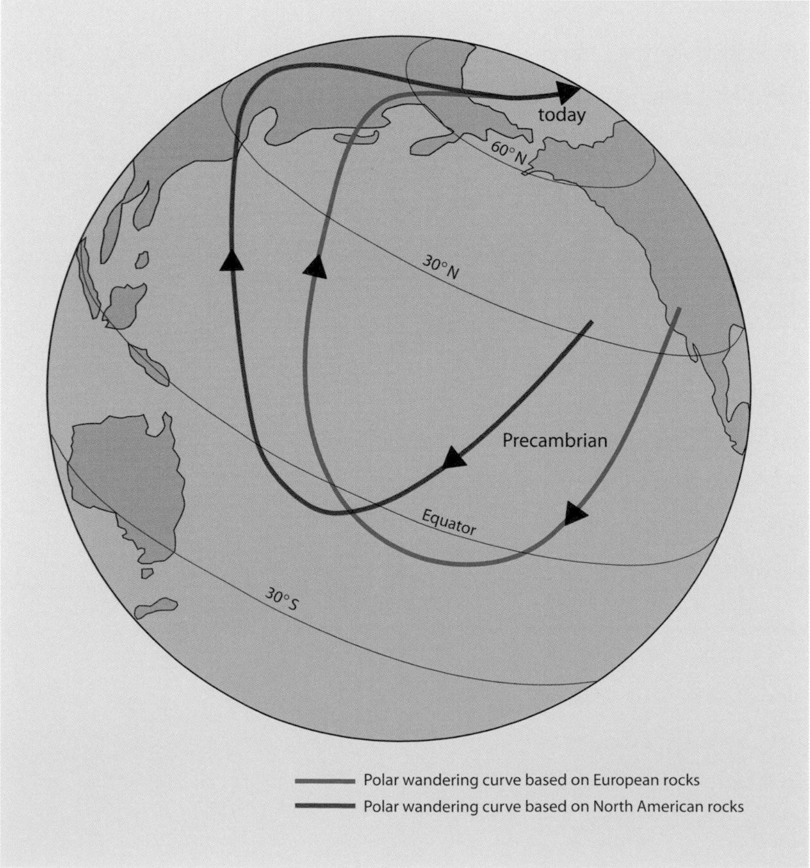

When data from other continents were compared with the European information, completely different polar wandering curves were derived for each continent. Since the Earth can have only one polar axis at any given time, it was the continents that must have drifted and not the poles. Furthermore, when the continents were moved back together, figuratively speaking, so that the polar wandering curves for the different continents merged, geological features and fossil distributions tended to match. Wegener had been vindicated.

Polar wandering curves are not the only important result of paleomagnetic studies. For reasons that are still not clearly understood, the Earth's magnetic field reverses from time to time. Magnetically, the North Pole becomes the South Pole and the South Pole becomes the North Pole. Reversals occur at irregular intervals, but happen on average about every half-million years. We know this from magnetic records in mainly volcanic rocks. As well as taking on the prevailing magnetic alignment, iron-containing minerals take on the existing magnetic orientation as they crystallize. They thus record "normal polarity" during intervals when the Earth's present magnetic North Pole is in the north, and "reverse polarity" when it is in the south. In the early 1960s, magnetic readings taken across the Mid-Atlantic Ridge south of Iceland revealed strips of crust parallel to the ridge with alternating normal and reversed polarity. And, centred on the ridge, there was a striking mirror-like symmetry of these strips.

By 1960, marine geologists had discovered that the Mid-Atlantic Ridge was one of an interconnected and branching series of ridges that together extend for a total length of 65,000 kilometres beneath the oceans. They had also recognized the youthful nature of the oceanic crust. To explain these observations, the American marine geologist Harry Hess proposed that the ocean floor is in motion, with molten rock welling up at the mid-oceanic ridges and then spreading outward across the ocean floor with time. As the new sea floor spreads, still more molten rock wells up to fill the space. The magnetic strips, which increase in age away from the ridge, are confirmation of Hess's idea of "sea-floor spreading".

A scientific revolution had been born, one as exhilarating and important to geology as Darwin's theories had been to biology and Einstein's theories to physics. Indeed, the 1960s were amazing years for geology. At the beginning of the decade, the mechanisms of continental drift were only vaguely understood. By the end of that decade, the concept of continental drift had evolved into the theory of plate tectonics, which has withstood the test of thousands of observations.

Plate Tectonics

If ocean floor is being created continuously at oceanic ridges, either oceanic crust is continuously being destroyed somewhere else, or the Earth is continuously expanding. The scenario of an expanding Earth has been dismissed for various compelling reasons. For example, if the Earth were smaller during the age of dinosaurs (but with the same mass, and thus denser), the law of gravity dictates that animals of the same size would have weighed considerably more than they do now. Dinosaurs would probably have collapsed under their own weight. However, we know that dinosaurs were a thriving group of animals, showing that the Earth has remained the same size. Hence, we can conclude that oceanic crust must continually be destroyed as well as created. In 1960, Hess suggested that deep-sea trenches, such as those around the modern Pacific Ocean, may be where oceanic crust is drawn down into the Earth's interior. Countless observations since 1960 have shown that Hess was correct: the oceanic crust is like a conveyor belt, with new crust rising at the oceanic ridges and sinking, or "subducting", at the trenches. This also explains why oceanic crust is so young.

Oceanic crust, which is relatively thin but heavy, can be subducted beneath relatively thick but light continental crust. For example, Pacific Ocean crust is now being subducted beneath South America. Or oceanic crust can be subducted beneath another section of oceanic crust, as is occurring today off Barbados. There, the Atlantic crust is diving below the Caribbean at an average rate of 2 to 4 centimetres per year. There are several consequences of this subduction of oceanic crust into the mantle. Earthquakes occur as the subducting slab becomes wedged under the overlying crust. As the slab descends, it is warmed, and the contained water

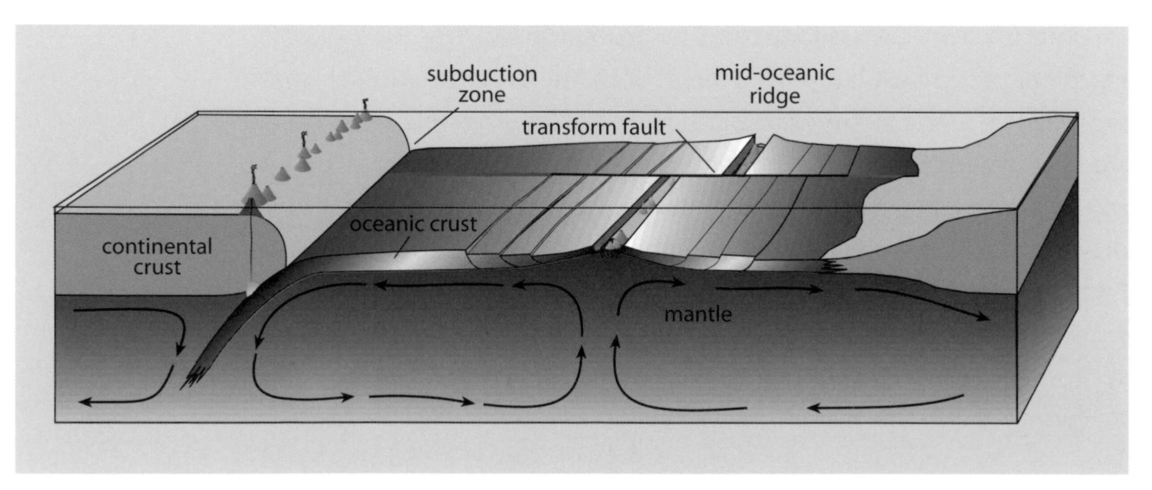

Relationships between the three types of tectonic plate boundaries, the mantle, and the two types of crust. The movement of tectonic plates is thought to be driven by convection currents within the mantle, as indicated by the arrows.

boils off and rises. This water lowers the temperature required to melt rocks in the mantle in the same way that ice will melt when salt is added. The molten magma thus generated rises to the surface. As it rises, the magma can collect in large reservoirs (or "plutons") in the crust. The granites that form highland areas in many parts of the Maritimes were formed in such plutons. Magma that reaches the surface forms chains of volcanoes, such as those of Japan.

The view along the main fracture of the Cobequid-Chedabucto Fault System near Parrsboro, NS: the road marks the fault. This fault system represents the former plate boundary between two small, ancient continents. It was once probably as active as the San Andreas Fault in California is today.

Subduction does not occur on all ocean margins. For example, the crust of the Atlantic Ocean is relatively young, and subduction occurs only locally, as in the Caribbean region mentioned above. When the continental parts of two plates collide, neither is heavy nor thin enough to be subducted. So, under colossal force, the margins of both can be pushed up to produce some of the highest mountains. For example, the Himalaya are still rising because of the collision between the Indian and Eurasian plates.

We have so far discussed two types of major crustal features marking plate boundaries: mid-oceanic ridges, where new crust is being formed by an upwelling of mantle material; and subduction zones, where oceanic crust is being destroyed. There is a third type: transform faults, where plates are grinding past one another. No crust is formed or lost at transform faults. The San Andreas Fault in California is a modern, active example of such a fault. The Cobequid-Chedabucto Fault System that runs across Nova Scotia is an ancient example, which thankfully is not still active.

Mid-oceanic ridges, subduction zones and transform faults divide the Earth's crust up into "tectonic plates", with the present-day Maritimes located on the North American plate. Most earthquakes and volcanoes lie along or near plate boundaries, confirming their status as the most active zones of a dynamic Earth. The study of plates, their history and movements is the focus of plate tectonics. Convection currents within the Earth's mantle probably control plate movements.

One more concept—that of an "orogen"—needs to be explained in this section. Where two plates converge, either at a subducting or a transform margin, the rocks become deformed and are thrust up as mountain chains. The Appalachians, which run through Atlantic Canada and the northeastern United States, are an ancient example; the Rockies and the Himalaya are more recent examples. All mountain chains are progressively worn down over time by erosion. Younger mountain chains are usually higher because they are being actively uplifted by plate tectonics. Older mountain belts are lower because they are no longer active. Indeed, many such belts are so eroded that they are no longer truly mountainous, and the main evidence for past cataclysms may be highly deformed rocks. So the term orogen, or "orogenic belt", is used for all such belts, presently mountainous or not, formed by convergent plate tectonic processes. The complex plate tectonic interactions responsible for an orogen is thus an "orogeny", or a mountain-building event.

Although the Maritime Provinces are today situated in the interior of a relatively stable plate, we know from their rocks that the three provinces share a tumultuous geological history. In addition to the Cobequid-Chedabucto Fault System, there have been several major volcanic episodes. One, for example, is represented by the lava flows (basalts) of New Brunswick's Grand Manan Island, and by those of North Mountain, a prominent north-bounding ridge to Nova Scotia's Annapolis Valley. And, of course, there are the Appalachians themselves, once possibly as high and majestic as the modern Himalaya. The story of the dynamic Maritimes is a main theme of this book.

Rocks—Nuggets of Information

To geologists, rocks are both time capsules and history books. They can tell us about past environments, climates, plate movements and conditions deep inside the Earth—indeed all kinds of interesting things. To tap this

The rock cycle

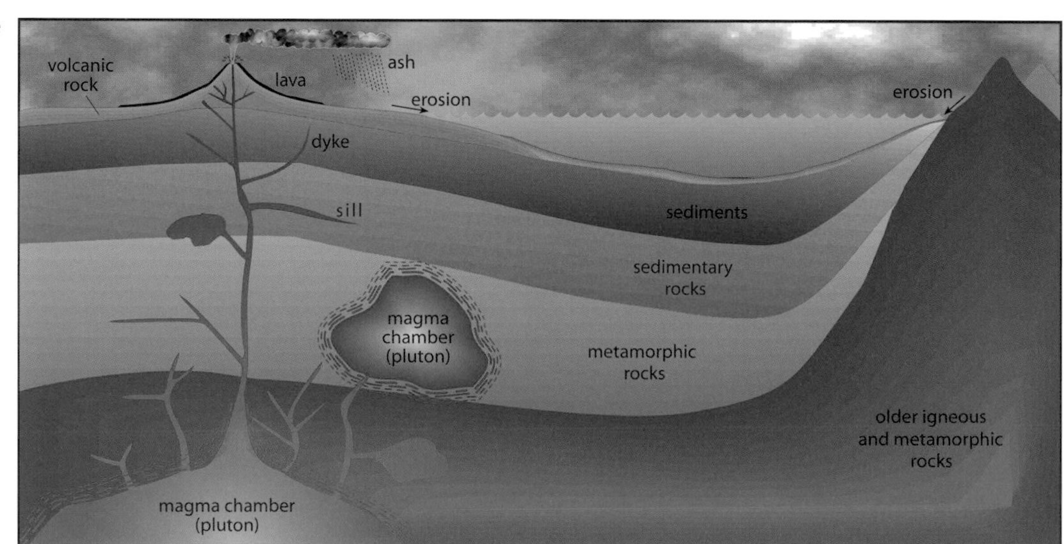

wealth of information, we need to know more about the three types of rocks: igneous, sedimentary and metamorphic. These types can be thought of as "fire rocks", "water rocks" and "altered rocks", and together they make up a dynamic "rock cycle".

Igneous rocks form from the cooling and crystallization of molten rock material called "magma". Because the Earth was very hot during its early history, the first rocks to form on this planet must have been igneous. Since then, magmas have cooled and crystallized quickly at the surface as fine-grained volcanic (or "extrusive") rocks, or slowly underground as coarser-grained "intrusive" rocks.

Examples of igneous rocks: two varieties of granite, pink and white, used as a building stone. St. Mary's Basilica, Halifax, NS.

Volcanic rocks vary depending on the type and volume of eruption. They are formed from lava flows or from "pyroclastics" (literally "fire fragments") such as ash and volcanic bombs. Intrusive rocks take many forms. They may solidify many kilometres below the surface, in bodies known as "plutons", where magma cools extremely slowly to produce coarsely crystalline rock such as granite. Magma can move from the plutons to the surface in pipes or sheets, and can also solidify in these forms. The solidified sheets are called "dykes" if they cut across layers of rock or through unlayered rocks, and "sills" if they are parallel to the rock layers.

A mafic dyke, cutting through other igneous rocks at Kennington Cove, near Louisbourg, NS.

Rocks are made up of minerals, and it is these that give rocks many of their properties. Colour, for example, is normally a good clue to the mineral composition of an igneous rock. Dark-coloured igneous rocks are called "mafic" and are rich in dark minerals such as pyroxene, biotite and olivine. These minerals contain high concentrations of iron, magnesium and calcium. An example of a dark-coloured igneous rock is basalt, the volcanic rock found along North Mountain, NS and on Grand Manan Island, NB. Light-coloured igneous rocks are called "felsic"; they contain high concentrations of sodium, potassium, aluminum and silica, and are rich in light-coloured minerals, such as feldspar and quartz. Examples of light-coloured igneous rocks are rhyolite, a volcanic rock found on the flanks of Sugarloaf Mountain at Campbellton, NB; and granite, a plutonic rock found at Peggys Cove, NS and Pabineau Falls, NB.

Sedimentary rocks originated as sediments and are usually found in layers, or "beds", collectively known as "strata". Rain, wind and waves caused the original igneous rocks—and later other kinds of rocks—to weather and erode, producing "clastic" sediments such as gravel, sand, silt and clay. These clastic sediments are washed or blown away and are eventually redeposited in rivers, lakes or the sea or as sand dunes in deserts. The beaches that are such a delightful part of the Maritimes are made up of sed-

The power of erosion: with the help of water-borne material such as sand and pebbles, the Nepisiguit River at Pabineau Falls, near Bathurst, NB, has cut into granite, one of the hardest of rocks. Note the light colour of the granite, which is typical of felsic igneous rocks.

An example of sediment transport and deposition by river and sea. Sand deposited at the mouth of Schooner Creek, near Cable Head, PEI, is also being moved and redeposited by waves of the Gulf of St. Lawrence.

iments produced by the erosion of older rocks. The layers of clastic sediment that accumulate, sometimes up to several kilometres thick, are changed into rock through compaction by overlying sediments and through cementation by chemicals in percolating ground water. Thus, gravels become conglomerates, sands become sandstones, and muds and clays become mudstones or shales. Some sedimentary rocks are formed by precipitation of salts from sea water or salt lakes, especially under hot, arid conditions. These rocks are chemical, rather than clastic, in origin, and are called "evaporites". They include salt and gypsum.

Limestone is also a chemical sedimentary rock, formed mainly of calcium carbonate. This white compound is familiar to many people living in areas underlain by limestone because, when dissolved in high concentra-

A conglomerate at St. Andrews, NB, made of pebbles and boulders in a natural sandy clay cement. The sediment was deposited by an ancient river about 370 million years ago.

White anhydrite and dark limestone deposits at Cheverie, NS; these deposits were formed in a shallow tropical sea, about 335 million years ago. Anhydrite is the mineral that gypsum converts to if it loses its water content after burial. Hence, the anhydrite shown here was deposited originally as gypsum.

tions, calcium carbonate tends to coat the inside of objects such as pipes and kettles. Similarly, calcium carbonate can be directly precipitated from sea water, although this occurs only in very warm climates. Usually, in nature, organisms make shells from the calcium carbonate in the water. If the water is warm and these organisms thrive, they may produce enough shells to form entire rock units. Such limestones, made up of shells, are called "coquinas". Some limestones are the remains of reefs that built up in ancient tropical seas.

Metasedimentary rocks at Blue Rocks, NS. The light and dark banding represent original sedimentary layers. The lineation marked by cracks in the rock represents cleavage, a metamorphic feature resulting from compression.

Sedimentary rocks commonly provide clues about how they were deposited. Such clues include ripples, dunes, raindrop impressions and mud cracks that formed when the sediment was deposited. Watch for these structures the next time you clamber over sedimentary rocks along the shore.

The third and final rock type is metamorphic. The word "metamorphic" means "changed form". Metamorphic rocks are produced by the effects of high temperature and pressure acting on pre-existing rock—either igneous, sedimentary, or even metamorphic—at some depth below the Earth's surface. Considerable erosion must take place before metamorphic rocks become exposed at the surface. Some of the metamorphic rock that now outcrop in southwestern Nova Scotia, for example, were formed at depths of 10 kilometres or more. During metamorphism, the minerals and textures of the previous rock change, producing new rock types, such as slate and schist from shale, marble from limestone, and quartzite or metasandstone from sandstone. (Geologists sometimes add the prefix "meta" to a rock name to indicate a metamorphic equivalent, such as "metasandstone".)

Rocks can be weakly to strongly metamorphosed, depending upon the intensity of the temperatures and pressures involved. "Low-grade" metamorphic rocks, such as slate and metasandstone, result from weak metamorphism and may still possess some original features, such as bedding and ripple marks. Low-grade metamorphic rocks develop a feature called "cleavage", whereby minerals—especially micas with their platy crystals—become aligned at right angles to the direction of compression. This is why schists (mica-rich rocks with well-developed cleavage) and slates split so easily and usually not along the bedding of the original sedimentary rock.

"High-grade" metamorphic rocks (those that have been strongly metamorphosed) are so intensely altered that it is usually difficult or impossible

Folded sedimentary strata exposed along the shoreline near Cheverie, NS.

to determine the original rock type. Gneiss is a common high-grade metamorphic rock, recognizable by its banded texture. The bands indicate where minerals have become segregated into layers of different composition during metamorphism. Like cleavage, the banding superficially resembles sedimentary bedding.

Examples of metamorphic rocks in the Maritimes include marble in road cuts around Saint John, NB, and along the Cabot Trail on Cape Breton Island. And rocks along the South Shore of Nova Scotia are mainly slates, schists and metasandstones.

Folds, Faults and Unconformities

Sediments are deposited as horizontal layers that gradually compact to form sedimentary rocks. Some areas, particularly those away from active plate boundaries, have a "layer cake" style of geology, in which all the beds remain flat-lying. Rocks of the Grand Canyon region in the American southwest show this style of geology. In such areas, individual horizontal layers can be traced hundreds—even thousands—of kilometres. Closer to active plate boundaries, especially in places where the crust is being compressed, strata are commonly deformed. Depending on physical conditions, such as temperature and pressure within the crust, rock layers may either bend to form "folds" or break to form "faults". If folded, the strata can be bent downward to form a "syncline" or upward to form an "anticline". Folding can be so intense that strata can be turned upside down, or "inverted". It can be on any scale, from microscopic to regional. On a regional scale, folding is seen as tilted strata in field exposures .

Folded sedimentary rocks at Rainy Cove, near Walton, NS. These rocks were formed as lake sediments about 360 million years ago and were folded as two continental blocks collided.

When there is compression (forces pushing together) or tension (forces pulling apart), rocks can break and move, causing earthquakes. Faults are the surfaces along which rocks break and are displaced. In compressional situations, "reverse faults" and "thrust faults" predominate because they heave one crustal block on top of another, thus shortening the crust. In tensional situations, "normal faults" tend to occur, moving one crustal block away from another and thus lengthening the crust. And where horizontal forces predominate, "strike-slip faults" occur, with one crustal block moving horizontally past another. Transform faults, already discussed as one of the boundary types between tectonic plates, are strike-slip faults.

Folding and faulting are associated with uplift of mountains over millions of years. These mountains may be eroded and planed off, and eventually a fresh sequence of sedimentary layers may be deposited on top. When these new sediments compact to become sedimentary rocks, they contrast with the older rocks because they are not deformed. The younger rocks remain less deformed even after a new round of crustal dynamics. In other words, the shape of their layering will not be the same as the layering of the older rocks beneath: they will not conform. We call the surface that divides these two suites of rocks an "unconformity".

Fault, trending diagonally from top right to bottom centre, in 200 million-year-old basalts at Five Islands, NS.

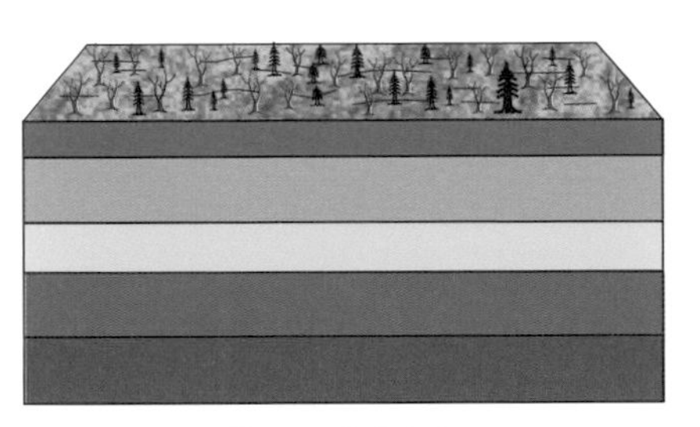

"Layer cake" strata

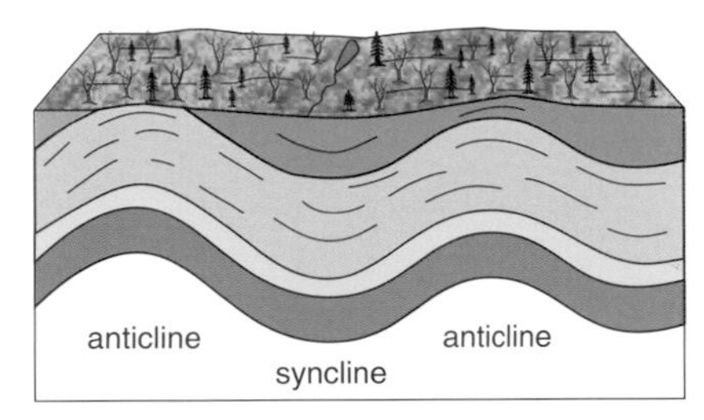

Simple folds

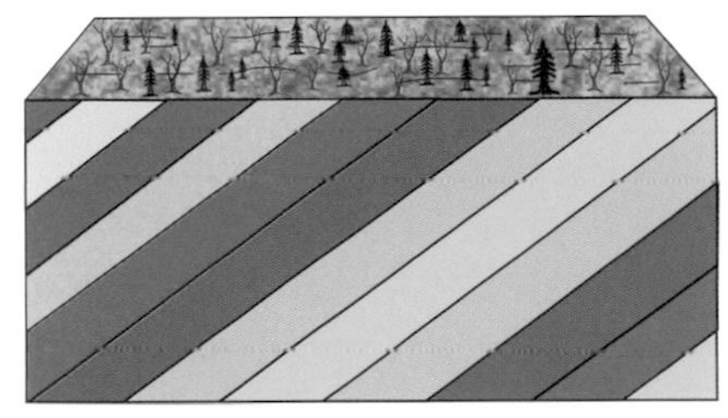

Tilted strata

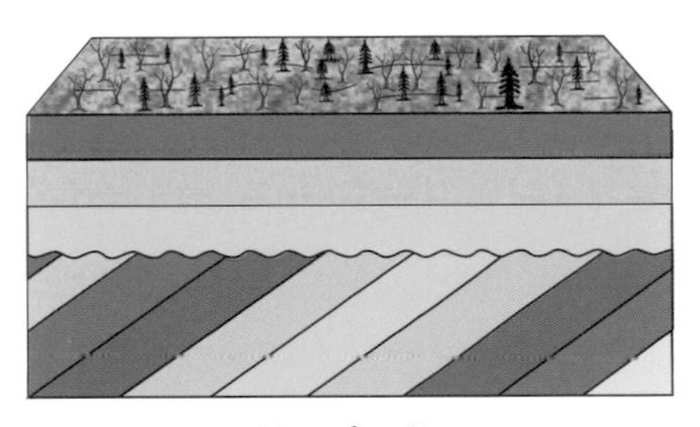

Unconformity

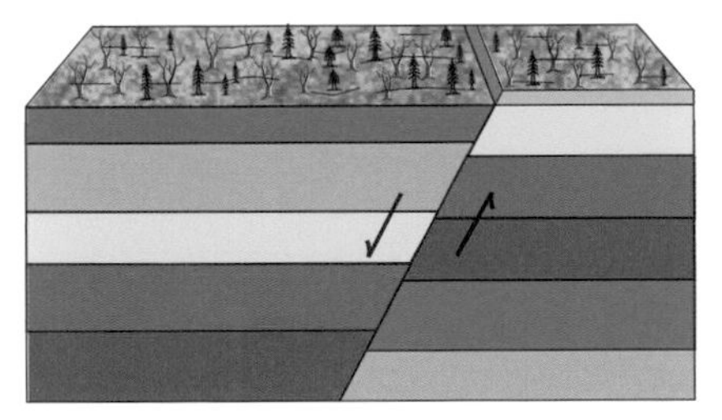

Normal fault

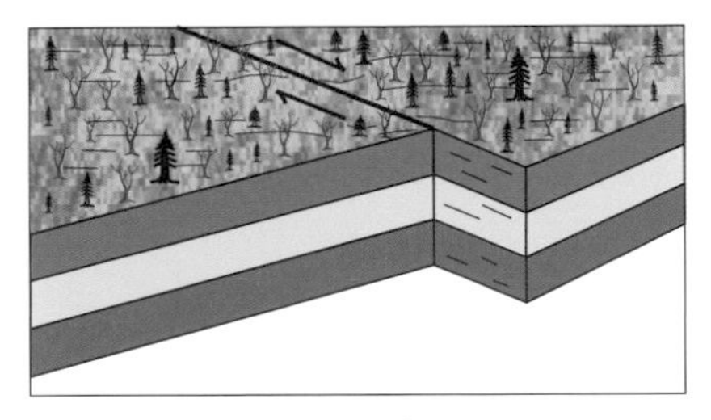

Strike-slip fault

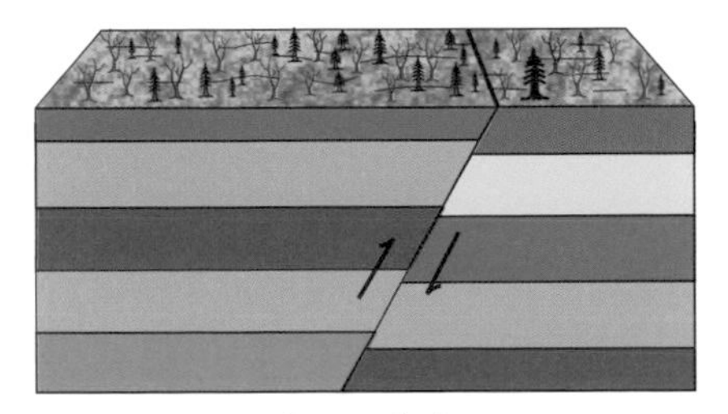

Reverse fault

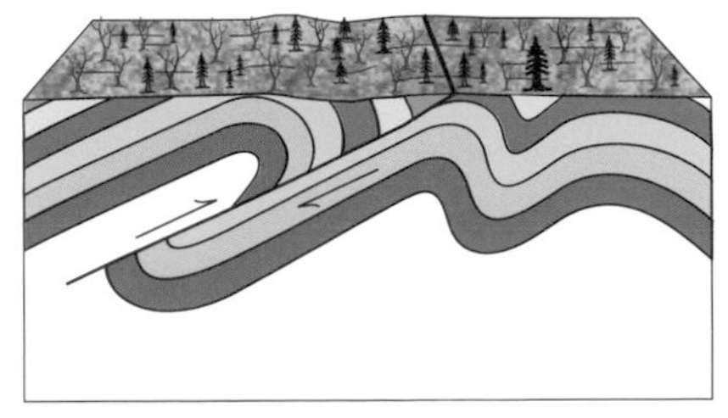

Thrust fault

A nodular "fossil" soil (paleosol) in about 250 million-year-old sedimentary rocks at St. Martins, NB. Such nodular paleosols are formed in warm climates with pronounced seasons.

Oceans, Plate Tectonics and a Constantly Changing Climate

By studying the rocks, we can see how climate has changed over time. For example, great salt pans covered the Maritimes about 335 million years ago, forming our salt and gypsum deposits in an arid climate. A wetter climate and tropical forests about 310 million years ago gave the region its coal deposits. The calcareous nodular fossil soil (paleosol) at St. Martins, NB, suggests a warm, relatively dry but seasonal climate about 250 million years ago. Similar soils occur today in many parts of the western United States and in southern Africa. Warm, dry conditions also prevailed about 200 million years ago, when thick salt deposits accumulated on the Scotian margin (today's continental shelf). In the past 2 million years, huge ice sheets covered the area. There is also evidence for earlier freezes, as shown by features in sedimentary rocks about 440 million years old at Cape St. Mary, NS.

Marine sedimentary rocks containing "dropstones", at Cape St. Mary, NS. Dropstones are pebbles and boulders dropped onto the sea floor from melting icebergs or an ice-shelf—in this case, about 440 million years ago.

What has caused such dramatic shifts in climate? The most obvious reason is changes in continental positions. Continental drift, driven by plate tectonics, exerts a slow but profound influence on climate. Although the continents on their plates move at about the same rate as fingernails grow, this movement adds up over millions of years, and explains how parts of the Maritimes, once in the southern hemisphere, crossed the tropics on their way north to their present positions. But the movement of continents alone does not fully explain the range of climatic change. The Earth's climate seems to go through relatively warm periods ("greenhouse" intervals), alternating with relatively cold spells ("icehouse" intervals). The present Ice Age can be thought of as an icehouse interval. And there is good evidence for major icehouse episodes in the past, including one that affected today's southern continents at the time when coal deposits were forming in the Maritimes. To understand the alternation between greenhouse and icehouse conditions, we need to understand the oceans.

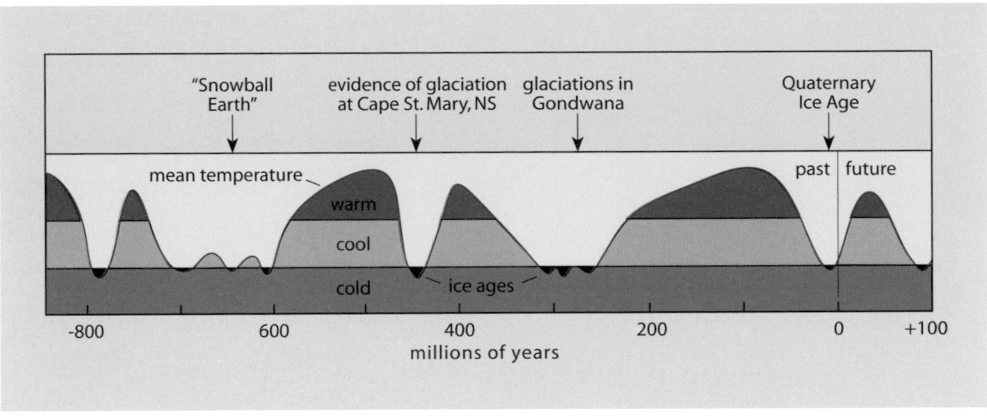

Changes in climate from icehouse to greenhouse over the last 800 million years, with a prediction for the future.

The oceans, as the atmosphere, have been likened to a great engine driven by energy from the sun. They can be thought of as vast, irregular containers with two layers of sea water: a thick, cold, lower layer and a thin, warmer, lighter, surface layer. Currents in the surface layer today include the Gulf Stream and the Labrador Current.

The main surface currents are driven by prevailing winds, and both surface currents and winds are deflected to the west by the Earth's rotation. This combination of factors drives the surface layer into a system of huge current gyres largely controlled by the location of major continents. Another example of how plate tectonics changes climate occurs when huge mountain ranges—such as the modern Himalaya and the ancient Appalachians—form through collision of drifting continents. Such mountain chains produce large areas of high land, leading to colder temperatures, increased snow accumulation, and changes to global patterns of air circulation.

Volcanic activity, which we mentioned earlier as being intimately linked to plate tectonics, may lead to global temperature changes through the ejection of huge volumes of ash into the atmosphere. The 1815 eruption of the Indonesian volcano Tambora was one of the biggest in historic times. There were reports of remarkable sunsets as far away as England for six months after the event. There are no photographs from that time, but these sunsets perhaps inspired some of the best works of the English painter J.M.W. Turner. The weather following the eruption of Tambora was so bad that 1816 was called "the year without a summer". And we know that the Tambora eruption was only a small event compared to many in geological history.

However, climate does not necessarily grow colder in association with volcanic events. Increased volcanic activity accompanying continental break-up produces more carbon dioxide that, in the long term, may cause a greenhouse effect and a warmer global climate. But in the short term, as shown by the Tambora eruption, it can be another story.

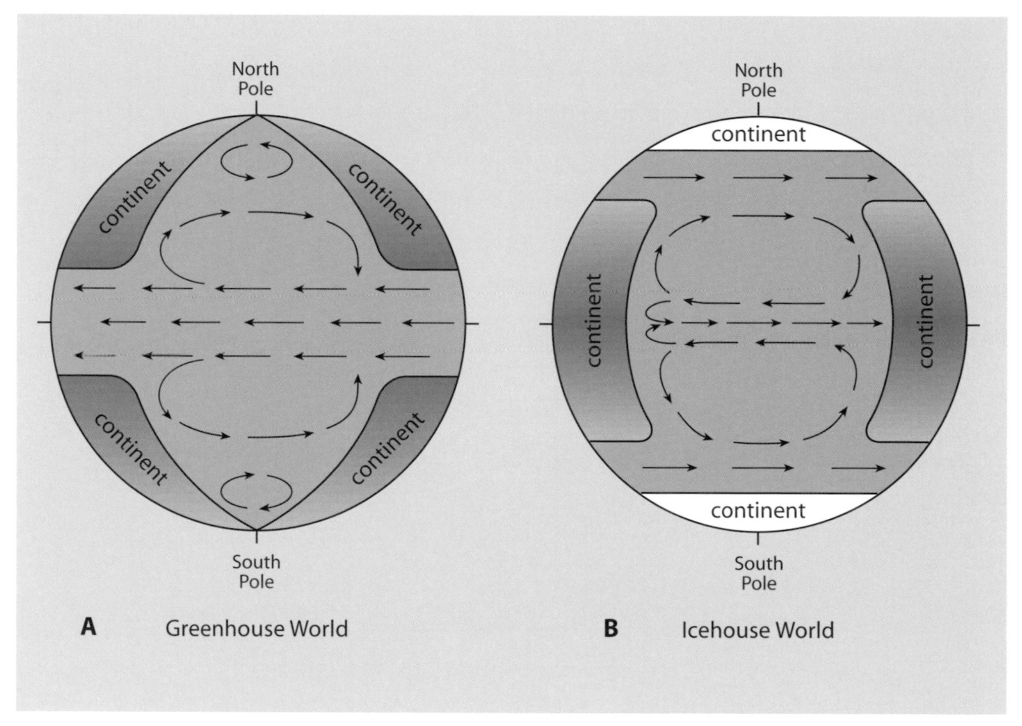

A. In an ideal greenhouse world, equatorial currents encircle the globe and so become much warmer. There is also complete exchange between tropical and polar waters. This combination leads to a planet that is too warm for polar ice caps to form.

B. In an ideal icehouse world, there are continents at the poles, and any currents encircling the globe tend to be polar rather than equatorial. This limits heat exchange between tropical and polar regions, promoting the formation of polar ice caps.

From Crust to Core: A 6,300 Kilometre Journey

When we gaze out from Maritime shorelines across the vast Atlantic Ocean, it is difficult to imagine the far coast, about 5,000 kilometres away. It is even more difficult to visualize the centre of the Earth, 6,300 kilometres beneath our feet. How can we find out what makes up this huge spinning planet of ours? The best clues are from earthquakes.

During an earthquake, the Earth literally quakes or vibrates. All earthquakes (including those induced artificially by bombs or controlled explosions set off by geophysicists for the purpose of exploring the Earth's structure) create seismic waves, which are like sound waves and can pass through rock and water. These waves radiate outward and downward from the epicentre or explosion. From an analysis of seismic waves, we can gain clues about the Earth's structure and composition. This process is something like the way that medical doctors, during an ultrasound examination, can "see" an unborn baby in its mother's womb.

Early analyses from seismic waves showed that the Earth's interior was like a series of nested balls. At the centre is the solid inner core, with a diameter of about 2,740 kilometres and an unbelievably high temperature of 3,000°C. Because of its high density (about 14 times that of water), the inner core is believed to be formed of iron and nickel. The inner core is surrounded by a 2,000 kilometre-thick outer core, which must be liquid to account for the Earth's magnetic field.

The core is enclosed by the mantle, which is about 2,900 kilometres thick and which makes up more than 80 percent of the Earth's volume. The upper mantle is about 700 kilometres thick and is somewhat lighter than the lower mantle. Most of the mantle is thought to consist of peridotite, a dark-coloured, coarsely crystalline rock rich in iron and magnesium.

The outermost rock layer of the planet, the part we live on, is called the crust. Compared to the mantle and core, the crust is very thin (about one hundredth or less of the Earth's radius) and is relatively weak. It is generally thickest (about 60 kilometres) under mountain ranges, and thinnest (as little as a few kilometres) under the oceans. In fact, there are two types of crust: oceanic and continental. Oceanic crust is similar in composition to the mantle, consisting of mainly iron- and magnesium-rich rocks like basalt. Continental crust is richer in silica, having an overall composition much like that of

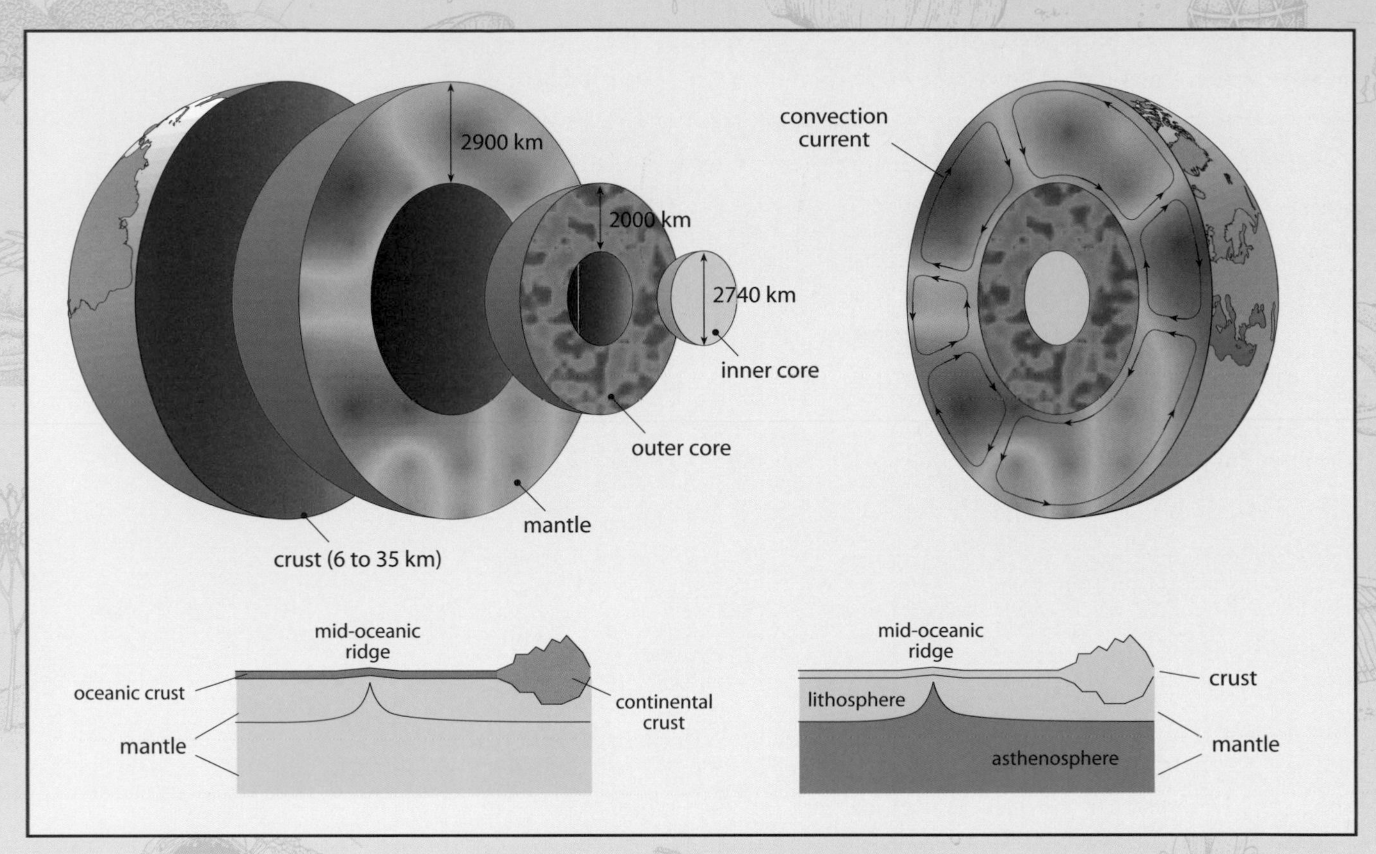

Structure of the Earth.

granite, and extends beneath the continents and continental shelves, beyond which it merges gradually with oceanic crust.

Although the crust consists mainly of igneous and metamorphic rocks, a small percentage (about 5 percent) is formed from sedimentary rocks. Sedimentary rocks are disproportionately represented at the surface, however, because that is where they are formed. Above the crust, there are three outer domains: the hydrosphere (areas of water such as oceans and lakes), the biosphere (all organisms), and the atmosphere.

The traditional division of the Earth into inner and outer core, inner and outer mantle, and crust is based largely on composition. In recent decades, geophysicists have developed a division based on strength and fluidity, this scheme complementing (rather than replacing) the traditional one. Strength is controlled by rock type, temperature and pressure. Generally, temperature is the most important factor, since hotter material flows more readily and thus appears weaker. In the new scheme, the crust and the top part of the upper mantle make up a strong, rigid outer layer called the "lithosphere". This layer varies in thickness from about 250 kilometres to nothing where deeper, hot material is flowing upward (generally at mid-oceanic rifts).

The weaker material underlying the lithosphere is called the "asthenosphere", which flows very slowly, a few centimetres per year. The driving force for this flow is the temperature difference between the top and bottom of the asthenosphere, which in turn causes convection cells to form. The asthenosphere is weakest right underneath the lithosphere, and much of the magma that is the source for volcanoes and plutons is generated there. The relative weakness of the asthenosphere allows the lithospheric plates to move around and interact in the process of plate tectonics.

To find out why distinct layers occur within the Earth, we need to consider the planet's origins. Earth apparently originated as a huge cloud of cosmic dust orbiting the sun. It grew by attracting other particles and gaseous material to itself. There may have been some layering in this dust cloud, with the denser material in the centre, but the separation into layers was caused mainly by high temperatures generated by radioactive decay. The melting that resulted from these high temperatures intensified the settling of heavier elements to form the core, while lighter fractions rose to the surface to form the crust.

But when were oceanic and continental crust formed? And have the proportions of these types of crust remained constant throughout the history of the Earth? The earliest-known rocks on Earth, dating from about 4 billion years ago, are metamorphic, but are derived from water-deposited sedimentary rocks. Thus, by that early time there must have been a hydrosphere, an atmosphere and at least some rocks of possible continental composition. Other old assemblages of rocks, from such areas as the Canadian Shield, seem to indicate that only about 5 to 10 percent of the present amount of continental crust existed around 3.5 billion years ago. Most continental crust probably formed between that time and about 500 million years ago, with the first larger continents developing about 2.5 billion years ago. Before that date, the Earth would have been internally warmer, resulting in more active and rapid creation and destruction of crust.

The first generally accepted evidence of true plate tectonics—with drifting, rifting and subduction—is from about 1.5 billion years ago, a long time after the Earth's beginning, but well timed in relation to the evolution of life on our planet. Eukaryotic cells (those cells with a nucleus, unlike bacterial cells) also arose at about that time. It is fascinating to speculate that these two major events in earth history may have been related.

Minerals: Earth's Building Blocks

Rocks are made up of one or more types of mineral. But what exactly is a mineral? To a geologist, a mineral is a naturally occurring inorganic material of a single chemical composition, usually in the form of crystals. Diamond, for example, is a mineral formed from only one element: carbon. But most minerals are formed from two or more elements that bond together to form compounds. An example is quartz, made up of the elements silicon and oxygen, which bond together as silicon dioxide. Minerals have characteristic crystal shapes and structures because the individual atoms that make them up (in the case of quartz, silicon and oxygen atoms) are arranged in a set pattern.

More than 3,000 minerals have already been described, and about 100 new ones are added each year. How can we identify different minerals? The chemical composition is the most definitive indicator, but this can be confirmed only through laboratory analyses. In studying hand specimens, we must rely on visual differences such as colour, streak, lustre, hardness and crystal shape.

The colour of a single mineral can be extremely variable. For example, quartz can be colourless and as clear as glass, in which case it is termed "rock crystal". But minute amounts of impurities (elements other than silicon and oxygen) produce dramatically different colours. Amethyst, one of the most beautiful varieties of quartz, owes its deep violet hue to the presence of iron. Traces of manganese or titanium impart a beautiful pink tint to rose quartz. And smoky quartz, the brown to grey variety, has been affected by natural radioac-

Cubic crystals of pyrite ("fool's gold"), set in volcanic ash, from New Brunswick.

Amethyst, the purple variety of quartz, with heulandite, a white-to-colourless zeolite mineral. This specimen is from Nova Scotia

Smoky quartz from Nova Scotia.

tivity. The most startling variety of quartz is multicoloured agate, the different coloured bands reflecting layers of microscopic quartz crystals. Agate eggs, which can be purchased from many rock and mineral shops, are breathtaking in their beauty. Although colours can be misleading, some minerals have a remarkably consistent hue. Malachite, a copper carbonate mineral, is always green. Azurite, another copper carbonate mineral, is always blue. And sulphur is yellow.

Multicoloured agate surrounding a core of purple amethyst, from Nova Scotia (the exact location is unknown).

The streak of a mineral is its colour when it is in powdered form. To determine this, geologists scratch a so-called "streak plate", a piece of unglazed porcelain, with the mineral. The mark left is the colour of the streak, which is constant for individual minerals. Thus, different oxides of iron have different streak colours: magnetite is black and hematite is brownish to cherry red. One drawback of this technique is that it usually does not help in identifying light-coloured or very hard minerals.

Lustre is the way in which the surface of a mineral reflects light. Those minerals that look like metal—for example, pyrite ("fool's gold")—are said to have a metallic lustre. Non-metallic lustres can be described as vitreous (glassy, like quartz), brilliant (like gems) and earthy (dull, like howlite).

Perhaps the most commonly used property of a mineral is its hardness, which is expressed as a value of 1 (softest) to 10 (hardest). This scale was the brainchild of the German mineralogist Frederick Mohs (1773-1839), and is thus called Mohs Scale. Mohs chose a representative mineral for each number: 1 is talc, 2 is gypsum, 3 is calcite, 4 is fluorite, 5 is apatite, 6 is orthoclase feldspar, 7 is quartz, 8 is topaz, 9 is corundum and 10 is diamond. Hence, quartz will scratch calcite but not topaz, and so on. Most rock and mineral stores sell hardness kits, but there are also quick and simple tests. A fingernail has a hardness of 2.5, a copper cent coin is 3, a knife made of steel is 5.5 and emery cloth is about 8. Hardness generally reflects the mineral's structure rather than its composition. For example, carbon forms one of the softest minerals (graphite, with a hardness of 1), as well as the hardest (diamond, with a hardness of 10), depending upon the arrangement of its atoms.

Howlite, a borate mineral found in association with gypsum in Nova Scotia.

The crystal shape often adds a final touch to the elegance of a mineral. This shape is simply an external expression of the orderly internal arrangement of the atoms. For example, pyrite occurs as cubic crystals, diamond is diamond-shaped, and quartz forms elongate prismatic crystals. Minerals of the zeolite group also have striking

Stilbite, a zeolite, from Partridge Island, near Parrsboro, NS. Stilbite is the provincial mineral of Nova Scotia.

crystal shapes. Zeolites are common in basalts (solidified lava flows) found around parts of the Bay of Fundy. They usually occur in holes, or "vesicles", that formed as gases escaped from the cooling lava. Seeping ground water later precipitated minerals, including zeolites, in these vesicles. The zeolite group of minerals includes natrolite, which has needle-like crystals that occur in radiating masses; mesolite, with its delicate fibre-like crystals; and stilbite, with platy crystals that sometimes occur in groups resembling old-fashioned haystacks. Some minerals, such as gypsum, can have several different crystal shapes.

Gems are special minerals. They have attractive colours, but to be truly considered a gem, a mineral must also be hard and uncommon. Most gems are cut from one of about 20 minerals. Thus blue sapphires and red rubies are both made of corundum, one of the most durable of minerals, with a hardness of 9.

The coarsely crystalline variety of gypsum known as selenite, found near Windsor, NS.

Although there are over 3,000 known minerals, it is perhaps surprising that there are not more, especially considering that there are 92 naturally occurring elements. But all elements are not found in equal abundance. The bulk of the Earth's crust is made up of only eight elements: oxygen, silicon, aluminum, iron, calcium, sodium, potassium and magnesium. Since oxygen and silicon together make up about three-quarters of the crust, silicates are by far the most common minerals. This explains why quartz, essentially pure silica, is so common. Another group of common minerals is the carbonates (with carbon and oxygen), which include calcite and dolomite. Calcite is the primary mineral in limestone and also forms many seashells. Other important groups are the sulphides, which contain sulphur (for example, the iron sulphide mineral pyrite); sulphates (for example, the calcium sulphate mineral gypsum); oxides (for example, the iron oxide mineral hematite); and halides (for example, the sodium chloride mineral halite, otherwise known as table salt).

The variety of gypsum known as "satin spar" because of its shiny, satin-like appearance. Satin spar is formed of fibrous crystals that are generally found in masses, as in this specimen from Nova Scotia.

Minerals are vital to the economy since they are found in nearly all manufactured products. Canada is an important producer of mineral resources such as copper, diamonds, gold, gypsum, nickel, potash, silver and sulphur. In the Maritimes, minerals of economic importance have been a major source of economic growth. This story is told in Chapter 8.

The evolving Maritimes: a depiction of events that occurred here during geological time. All animals and plants shown are based on fossils from the three provinces, and the scenery is also based on stages in the region's past.

CHAPTER 2
The Fourth Dimension

A Sense of Time

In 1650, James Ussher, Bishop of Armagh and a leading biblical scholar, published his book, *Annals of the Old Covenant from the First Origin of the World*. In this learned tome, Bishop Ussher pointed out that previous philosophers had abandoned any attempt to determine the time of the world's origin. Fortunately, his generation's knowledge of the Hebrew calendar, astronomy, the Bible and other records resolved this problem. The origin of the world, the Bishop and his acolytes calculated, was the night preceding Sunday the 23rd of October in the year 4004 BC. This would make the Earth about 6,000 years old. Although Bishop Ussher's work came to be ridiculed by later generations (especially by writers of geological textbooks), he was on the right track, indeed ahead of his time, in trying to establish a chronological history of the Earth.

Early humans did not measure time in minutes or seconds, but they did recognize seasons, the regularity of which is determined by the movement of the Earth through the solar system. They attached great importance to celestial events such as the summer solstice, when the Sun in the northern hemisphere is at its highest, and the winter solstice, when the Sun is at its lowest. This information was vital to their survival because it told them when to plant their crops and when to prepare for the rigours of winter.

But how did our ancestors determine the summer solstice? Early attempts may be preserved in such monuments as Stonehenge and the Egyptian pyramids. In time, calendars were devised to organize the days of the year. As with most human innovations, several calendar systems were developed, the most prevalent today being the Gregorian calendar, introduced by Pope Gregory XIII in 1582.

Alongside the development of calendars, there was the need for devices to measure the passage of time within each day. The first attempts involved primitive sundials, which measured the shadow of the Sun as cast by a stick–a method not overly practical in a place like the Maritimes, where a shadow cannot be guaranteed. Then came the water clocks of the ancient Egyptians, followed by increasingly sophisticated timepieces, until today we have the quartz watch, claimed to be accurate to within one second in every ten years. Our most precise timepiece is the atomic clock, accurate to one second every 10,000 years. This gives us the ability to measure the minuscule slowing of the Earth's rotation due to the tidal effect of the moon.

How do humans think about time? We are comfortable with the seconds, minutes and hours within each day and can easily translate these into the years of our lives. Historical time is more difficult to comprehend–for example, think about the 5,000 years since Egyptian civilization began. And to project further back in time to the last glacial episode, more than 10,000 years ago, is mind-boggling. Geological time, which involves millions and even billions of years, is almost inconceivable. The oldest rocks in the Maritimes are about one billion years old–but are still relatively young when compared with the Earth, which is now believed to be about 4.6 billion years old.

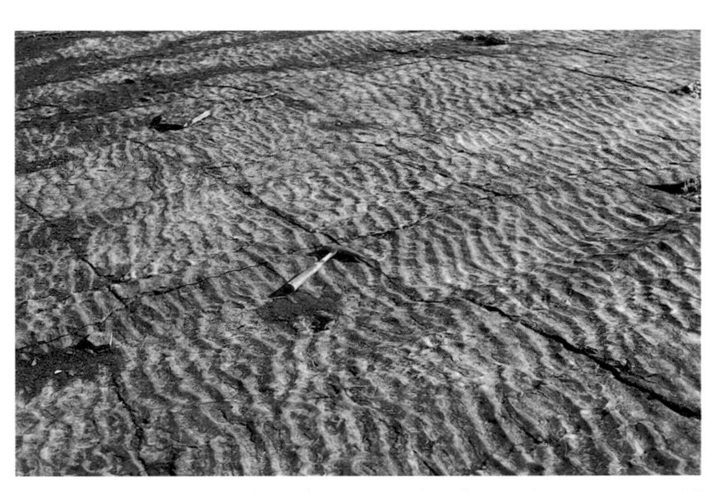

These two photographs demonstrate how similar processes can occur at different geological times. The photograph to the left shows modern ripple marks on the exposed tidal flats at Medford, NS. The other photograph also shows ripple marks, but these examples are preserved in sedimentary rock at Blue Beach, near Hantsport, NS. They formed on a lake shore about 360 million years ago and are exposed today because overlying rock has been removed by erosion.

Relatively Speaking

Because the concept of time is so hard to grasp, the recognition of the extent of geological time was a painfully slow process. Some rocks seem to be older than others because they are harder or more altered, but experience has taught geologists that such observations are notoriously unreliable as guides to time. The first real steps at gaining an understanding of the subject were based on the concept of relative time–the order in which events happened. For example, dinosaur bones are not found in the same rocks as human fossils, so it is logical to assume that the two are of different ages.

Geologists have developed some basic principles in interpreting rocks. One is the "principle of superposition", first realized through the geological studies of Niels Stensen (1638-1686), a Dane who is usually known by his Latin name, Steno. This principle states that in flat-lying sedimentary or volcanic rocks, the oldest layer is at the bottom and the youngest layer is at the top. In Chapter 1, we described unconformities, where more steeply tilted older rocks underlie less steeply tilted younger rocks. An

An illustration of the principle of superposition. The younger Jurassic lava flows at the top of the cliff in the background overlie older Triassic red sedimentary strata (with a thin white layer in between). This view is at Five Islands, NS.

Unconformity at Jacquet River, NB, with almost vertical sedimentary rocks below, and gently tilted sedimentary rocks above. The lower rocks are about 420 million years old, and the upper ones are about 290 million years old. The unconformity therefore represents a hiatus of about 130 million years. Note the uneven surface of the unconformity, which represents an ancient erosion surface upon which the upper sediments were deposited.

A potential unconformity in the making at Joggins, NS. Note the ridges formed by the tilted sedimentary rocks, about 310 million years old. If this surface were to become permanently buried by sediments on top of the beach sand and gravel, we would have an unconformity like that shown in the previous photograph.

unconformity represents an ancient erosion surface and, usually, a considerable amount of time–tens, perhaps hundreds, of millions of years.

The principle of superposition is one way in which the rocks themselves give clues about their relative age. However, fossils are the most important tools that geologists use in determining the relative age of rocks. The "principle of fossil succession" was developed through the work of several people, including William Smith (1769-1839), an English civil engineer who was helping to build canals during the early prerailway years of the Industrial Revolution. This principle states that different fossils characterize different layers of rock. Since observations show that dinosaur fossils consistently occur in lower rocks than human fossils, dinosaurs must be older than humans. This example is perhaps obvious, but the same reasoning can be applied to other, less familiar fossil organisms—for example, trilobites lived and became extinct before ammonites first appeared. Fossil groups such as trilobites and ammonites will be explained in Chapter 3.

Without having to use numerical ages, geologists thus began to recognize a sequence of rocks around the Earth, from oldest to youngest, and to identify the distinctive fossils that are found in each part of the sequence. Not every part of this global succession of rocks (the "geological column") is found in any one area. For example, rocks bordering parts of the Bay of Fundy record the beginning of the age of dinosaurs, but to find rocks representing the end of the dinosaur era, we have to go offshore and drill into the Scotian Shelf. To find rocks bearing fossils of the earliest humans, we have to go to Africa.

By applying principles such as those of superposition and fossil succession, geologists have been able to subdivide the Earth's history into a sequence of relatively short intervals, geologically speaking. These intervals have been given names and together form the geological timescale. The two basic divisions, or "eons", of the geological column are the Precambrian and the Phanerozoic. The Precambrian includes the oldest rocks and con-

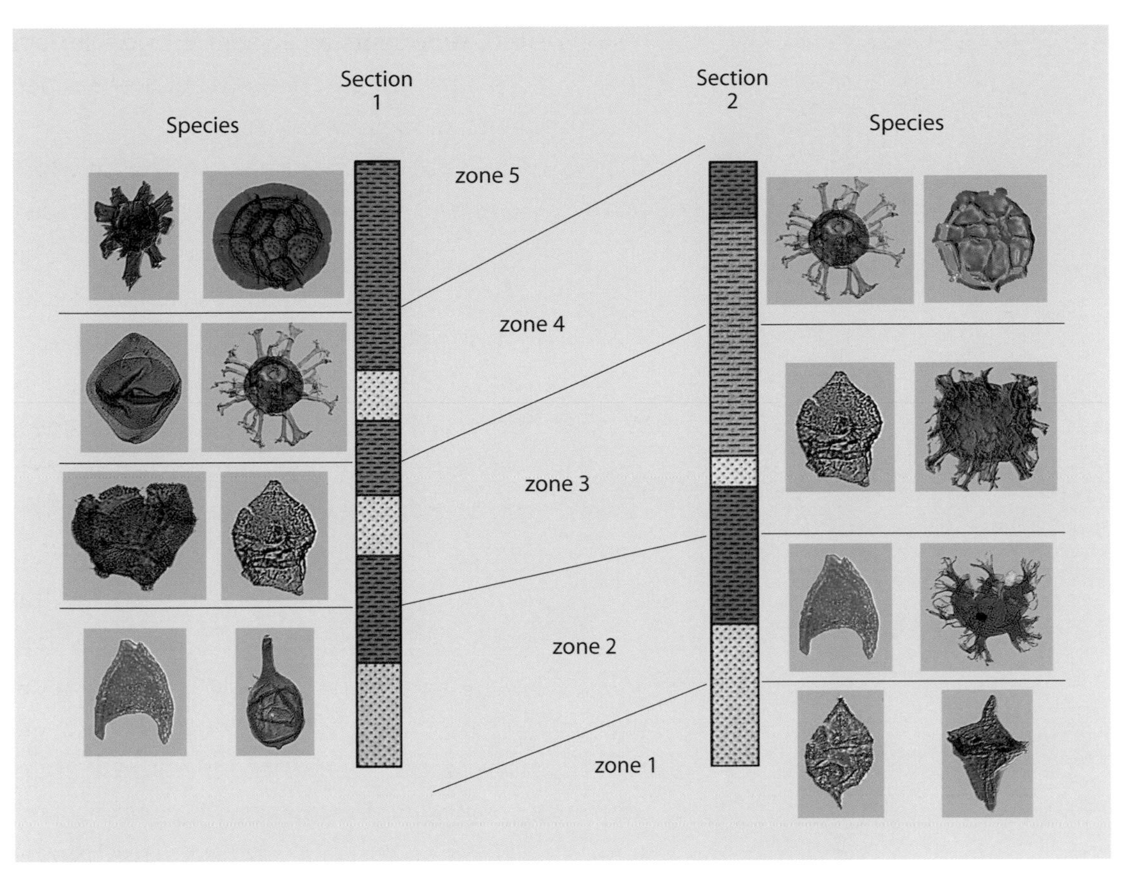

This diagram shows how fossils are used to link (or "correlate") rocks of the same age in different places. Not all fossils that occur at a certain time occur at all locations, but there is usually enough overlap to permit correlation. In most cases, the closer that two localities are, the easier it is to correlate them using fossils. The specimens shown here are fossilized single-celled organisms known as dinoflagellates. These "microfossils" are important tools in correlating the oil- and gas-bearing rocks on the Scotian Shelf. Modern dinoflagellates are a major group of marine plankton that are often the cause of "red tides" and paralytic shellfish poisoning.

tains no hard-shelled fossils, except for a few at the very end of that interval. The Phanerozoic, which means "visible life", began with the first appearance of common, hard-shelled fossils and continues to the present day. It is subdivided into three "eras". The oldest era is the Paleozoic, meaning "ancient life", and the youngest era is the Cenozoic, meaning "modern life". Between the Paleozoic and Cenozoic eras is the Mesozoic Era, or "middle life". These eras are further subdivided into units of time that geologists call "periods". The names of these periods, and of the rock "systems" representing the periods, have various origins. Some reflect the regions that were considered representative of the rocks of that age in the early days of geology. For example, the Devonian Period is named after the county of Devon in England. Others reflect a rock type dominating that part of the succession, such as the Carboniferous Period, named for the high carbon content of the widespread deposits of coal formed during this interval.

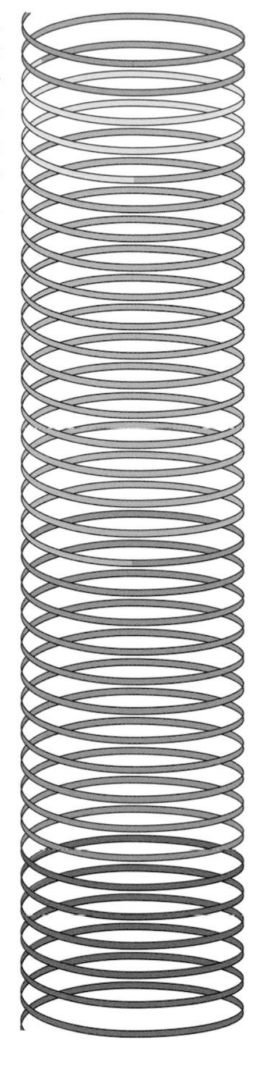

The geological column. The coiled spring to the left is to scale, with its colour scheme matching that of the geological column to give a better idea of the true relative length of each era.

Eon	Era	Period	millions of years	Event
Phanerozoic	Cenozoic (modern life)	Quaternary	1.8	First humans Glaciation Modern mammals evolve
		Tertiary: Neogene	24	
		Tertiary: Paleogene	65	First grass Greenland-North America split Evolutionary expansion of mammals
	Mesozoic (middle life)	Cretaceous	142	Peak, then extinction, of dinosaurs First flowering plants
		Jurassic	200	First birds; dinosaurs prominent Break-up of Pangea
		Triassic	248	First mammals and dinosaurs Mammal-like reptiles prominent
	Paleozoic (ancient life)	Permian	290	Reptiles dominant Maritimes in middle of Pangea
		Carboniferous	362	Major coal-forming swamps Southern hemisphere glaciers First reptiles and ferns Formation of Pangea
		Devonian	418	First trees Fish dominant First amphibians
		Silurian	443	Maritimes collage comes together First sharks
		Ordovician	495	Gondwana glaciation; invertebrates dominant First land plants
		Cambrian	545	First chordates and fish Diversification of invertebrates
Precambrian	Proterozoic		700 900 2500	First multicellular life Break-up of Rodinia Sexual reproduction; oldest eukaryotes Formation of Rodinia Oxygenation of atmosphere
	Archean		3400 4000	First prokaryotes Oldest known rocks
	pre-geological history of the earth		4600	Intense meteorite bombardment of earth Origin of atmosphere

Absolute Time

The concept of relative age has given us a geological column, with its eons, eras and periods, and these are a vital part of the story of geological time. To complete the story we need to add "absolute age", measured in years. In this way we can gain an appreciation for the pace of geological time. As mentioned earlier, Bishop Ussher made one of the first attempts to determine the numerical age of the Earth. But later thinkers doubted Ussher's estimation. One of the first to question such a young Earth was the French zoologist Compte de Buffon (1707-1788). He reasoned that the Earth had cooled from a molten mass to its present form at a predictable and gradual rate. Using experiments on the cooling time of molten metal balls of various sizes, he calculated that the Earth was 75,000 years old. This was a staggering difference from the age calculated using the Bible. But even more staggering estimates were to come.

A major step forward was made by Scottish geologist James Hutton (1726-1797), who proposed that rocks have been formed, deposited and eroded continuously over vast intervals of time by processes like those operating today. He concluded that the Earth showed "no vestige of a beginning, and no prospect of an end". Strictly speaking, Hutton was wrong. There was a beginning, billions of years ago, when the Earth formed. And there is prospect of an end, when the Sun dies, billions of years hence. But he was among those leading the way in visualizing vast amounts of time, a necessary precursor for the theory of evolution proposed by Charles Darwin in his *Origin of Species*, published in 1859. Darwin estimated that the dinosaurs lived 300 million years ago. But he and other natural scientists were dismayed when, soon after the publication of the *Origin of Species*, the famous but imperious English physicist, Lord Kelvin (1824-1907), "proved" how young the Earth really was. Starting with a molten earth and a rate of heat loss based on known temperature increases in deep mines, Kelvin concluded that the Earth was only 20 to 100 million years old. Unfortunately, Lord Kelvin was unaware of the existence of radioactivity, the Earth's internal heat engine. The discovery of radioactivity at the end of the nineteenth century made it feasible to again believe in vast amounts of time. And there was a bonus: radioactivity provided the means for measuring absolute time.

Absolute age dating is based on the radioactive decay of atoms within minerals. In 1903, Pierre and Marie Curie showed that, upon decay, radioactive atoms in minerals produce heat and change into other types of atoms. Radioactive decay is the process by which an unstable atomic nucleus of one element is transformed into a more stable nucleus of another element. For example, uranium 238 decays to lead 206 (the number is the atomic mass—the total number of protons and neutrons in each atom). In measuring the decay rate of elements, scientists use the term "half-life", which is the time taken for one half of the atoms of the unstable element to change to atoms of the stable element.

Scientists can derive the age of a rock from this type of calculation, first carried out by Ernest Rutherford (1871-1937) at McGill University in Montreal in the early twentieth century. Because igneous rocks are formed directly from magma, they give the most reliable ages. Sedimentary rocks cannot be dated directly. Volcanic fragments within sedimentary rocks can sometimes be dated, but the age will be that of the volcanic parent rock. Metamorphic rocks will give the date of their latest recrystallization–the last time that they were subjected to heat and pressure—and hence the age derived from a metamorphic rock will be a minimum age.

The results of absolute age dating can be summed up in one word–staggering! The oldest rocks still preserved on Earth have been shown to be about 4 billion years old. Using this evidence, along with information from meteorites and Moon rocks, most scientists now believe that the Earth is about 4.6 billion years old. That's old, but not as old as the universe, which astronomers believe to have formed 12 to 15 billion years ago. By dating igneous rocks, geologists have been able to give absolute ages to the various divisions of the geological column. The oldest eon, the Precambrian, ended about 545 million years ago. The Phanerozoic Eon, the time of shelly and skeletal fossils, covers the last 545 million years. Dinosaurs appeared about 230 million years ago and died out about 65 million years ago. Hominids (humans and their immediate ancestors) have been around for about the last 4 million years.

Radioactive elements with shorter half-lives, such as carbon 14, have helped to put an absolute-age framework onto younger parts of Earth history. We know, for example, that Nova Scotia was covered in ice about 15,000 BP (that's "Before Present", present meaning 1950) because of analyses of carbon in organic material from glacial sediment. Carbon 14 decays to stable carbon 12. Carbon-dating only works for about the last 50,000 years because of the relatively short half-life of carbon 14 (about 5,000 years). And to make resolution more precise in the most recent past, carbon 14 dates can be matched with tree-ring counts, which can be taken back to about 14,000 BP. In technical publications on the geology of the last 14,000 years, dates are usually not modified to take into account the tree-ring recalibrations, but in this book we have made the modifications.

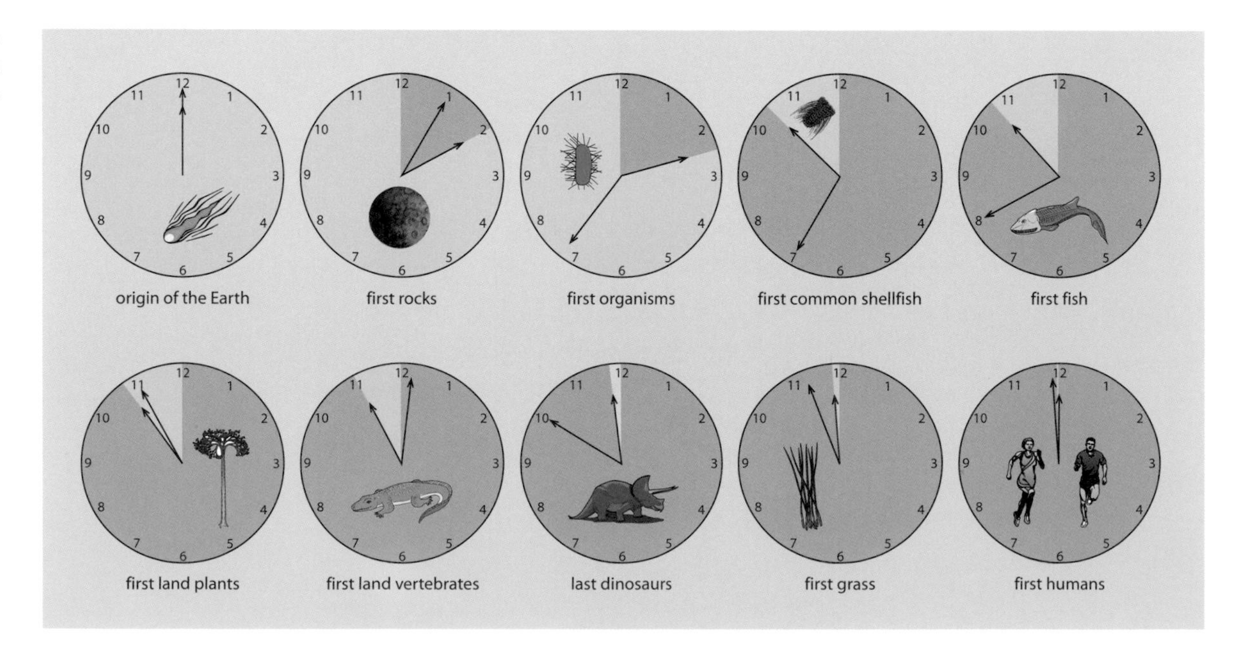

The history of the Earth as represented by a twelve-hour clock.

To help us understand geological time, and how little of it our species occupies, the history of the Earth can be represented by a 12-hour clock. If the Earth originated at midnight and if the present day is noon on the clock, the earliest known rocks occur at 2:05 a.m., the first known organisms at 2:36 a.m., the first common shellfish at 10:35 a.m., the first land vertebrates (amphibians) at 11:01 a.m. and the last dinosaurs at 11:50 a.m. Humans arrive only in the last minute before noon.

Today we are entering a new era of refinement in the precision with which we can determine ages. But appreciating the magnitude of the geological time scale remains a daunting exercise.

Ancient rocks from Saint John, NB, which were formed around the time of the Precambrian-Cambrian boundary, about 545 million years ago.

Modern sediment: dunes with ripple marks in the Minas Basin near Grand Pré, N.S.

People

Sir John William Dawson, author of "Acadian Geology"

John William Dawson (1820-1899) was born to Scottish parents in the village of Pictou, NS, and it was there that he first developed a love for the natural world. One day, he and some other schoolboys were shaping rocks into home-made pencils when young William noticed on his stone "a delicate tracing in black, of a leaf like that of a fern". Exploring further, "I found more fragments of leaves and soon had a little collection of them laid out on the shelf of a cupboard in which I kept my childish treasures". So began Dawson's life-long passion. He went on to write over 150 scientific articles, became the first Principal of McGill University, and was later knighted.

In 1841, Dawson was introduced to Sir Charles Lyell, the most influential of all Victorian geologists and author of the classic *Principles of Geology*. Lyell had come to visit the coal mines of Pictou during his first trip to North America. Eleven years later, Lyell returned, and he and Dawson struck out for the sea cliffs of Joggins, NS, where the pair made one of the most famous fossil discoveries in paleontology. In the stony cast of a once-hollow fossil tree, recently fallen from the cliff, they found the jumbled bones of what was until recently the oldest known reptile. Dawson later named this animal *Hylonomus lyelli*, meaning "Lyell's wood mouse". The puzzle of how a reptile skeleton could become fossilized in a once-hollow tree trunk continues to intrigue paleontologists.

Moment of discovery: Dawson, Lyell and an assistant (with hat) at Coal Mine Point, Joggins, discovering fossil vertebrates in the infilling of a fossil tree trunk.

From Joggins and elsewhere in the Maritimes (which Dawson nostalgically called Acadia) he made many other important discoveries of fossil life, great and small. These included fossil plants, trackways of lowly invertebrates, footprints, skeletons of reptiles and amphibians, millipedes and the earliest land snails. He had an uncanny ability to understand the ancient environments in which rocks had formed and to decipher their correct ages. Dawson died in 1899, but his greatest legacy is still with us—his classic book *Acadian Geology*, published in five editions from 1855 to 1892.

"Fall of the Economy River", near Economy, NS. From Dawson's *Acadian Geology*.

Abraham Gesner, Father of Oil Refining

Abraham Gesner (1797-1864), born near Cornwallis, NS, made major contributions to the geology of the Maritimes. As a young man, he became a keen student of rocks and minerals. But it was not until he had tried his hand at several other occupations that he became a geologist. After unsuccessful attempts at horse-trading and experimental farming, Gesner went to England to study medicine. He returned to Canada around 1826 and, for over a decade, practised medicine in Parrsboro, NS. During this time he studied the local cliffs and wrote a book on the geology of Nova Scotia, using his leisure time to study rocks.

In 1838, Gesner accepted the position of Provincial Geologist for New Brunswick, becoming the first Provincial Geologist in Canada. For four years he travelled throughout the province, mapped rocks, prospected mineral deposits and published five geological reports. Unfortunately, his conviction that coal deposits underlay a third of New Brunswick caused problems when miners failed to confirm this. Consequently, the provincial government withdrew his funding. Gesner then opened a museum in Saint John to try to earn a living by exhibiting his extensive collection of mainly geological specimens. This museum was the first of its kind in British North America and formed the nucleus of what became the New Brunswick Museum. The museum failed to make money, however, so Gesner returned to Nova Scotia to work as a doctor and to experiment, successfully, with distilling kerosene from oil shale. It was this discovery that brought him fame and, for a while, fortune.

A southern Nova Scotia scene, "Granite Hill and Lake near Saint Mary's River 1845". From Dawson's *Acadian Geology*.

In the late 1840s, a deposit of bitumen was discovered in Albert County in southeastern New Brunswick, and named "albertite". It was an ideal raw material for producing kerosene, so Gesner returned to New Brunswick and staked mineral claims on a property near Hillsborough. At that time, there was controversy over whether albertite was a mineral or a coal. This was an important issue, as others staked claims on the property to mine coal. Gesner lost that legal battle, as well as a battle over the patent to distil kerosene. He spent his later years working as an industrial chemist in the United States, but in 1863 (shortly before his death at the age of 66) he returned to Nova Scotia to become the first chemistry professor at Dalhousie University.

Throughout his varied career, Gesner wrote numerous scientific articles and several books, the most important being *A Practical Treatise on Coal, Petroleum and Other Distilled Oils*, published in 1861. This work has contributed to his recognition as the "father" of the oil refining industry. He is credited with several inventions, including an effective wood preservative, a process of asphalt paving for roads, briquettes of compressed coal dust, and a machine for insulating electric wire.

"Part of Cape Blomidon 1846", near Blomidon, NS. From Dawson's *Acadian Geology*.

George Frederic Matthew and The Steinhammer Club

George Frederic Matthew (1837-1923) was born in Saint John, NB. His formal education ended after grammar school and, at the age of fifteen, he entered public service in the Saint John Custom House. He stayed there most of his life, eventually becoming Chief Clerk and Surveyor. However, George Matthew had an extraordinary interest in geology and, in 1857, helped found the Steinhammer Club. This club was formed by a group of young men, none yet twenty years old, who were interested in learning about the rocks and fossils of the Saint John area. Sir William Dawson became interested in the activities of the Steinhammer Club, in particular the work of Matthew and his friend Fred Hartt (1840-1878), another famous New Brunswicker who later became director of the Geological Survey of Brazil and the first professor of geology at Cornell University in the United States. Matthew and Hartt provided information and specimens that Dawson used in his studies of New Brunswick. Dawson, as Canada's foremost geologist, likely exerted considerable influence on Matthew, who soon began assisting the Geological Survey of Canada. In addition to preparing map reports for the Survey, Matthew became the Canadian expert on Cambrian paleontology, and collections from British Columbia, Cape Breton and Newfoundland were sent to him for examination.

The varied geology of Saint John provided Matthew with a lifetime of research. As part of his extensive studies on Cambrian fossils, Matthew became one of the first paleontologists to recognize the early appearance of small shelly fossils. He also published the first scientific description of a Precambrian stromatolite. Many of the fossils collected by Matthew can still be found at the New Brunswick Museum.

In 1882, when the Royal Society of Canada selected charter members to represent the geological and biological sciences, George Matthew was among them. He received many honours, including, in 1923, the prestigious Murchison Medal of the Royal Geographical Society of London

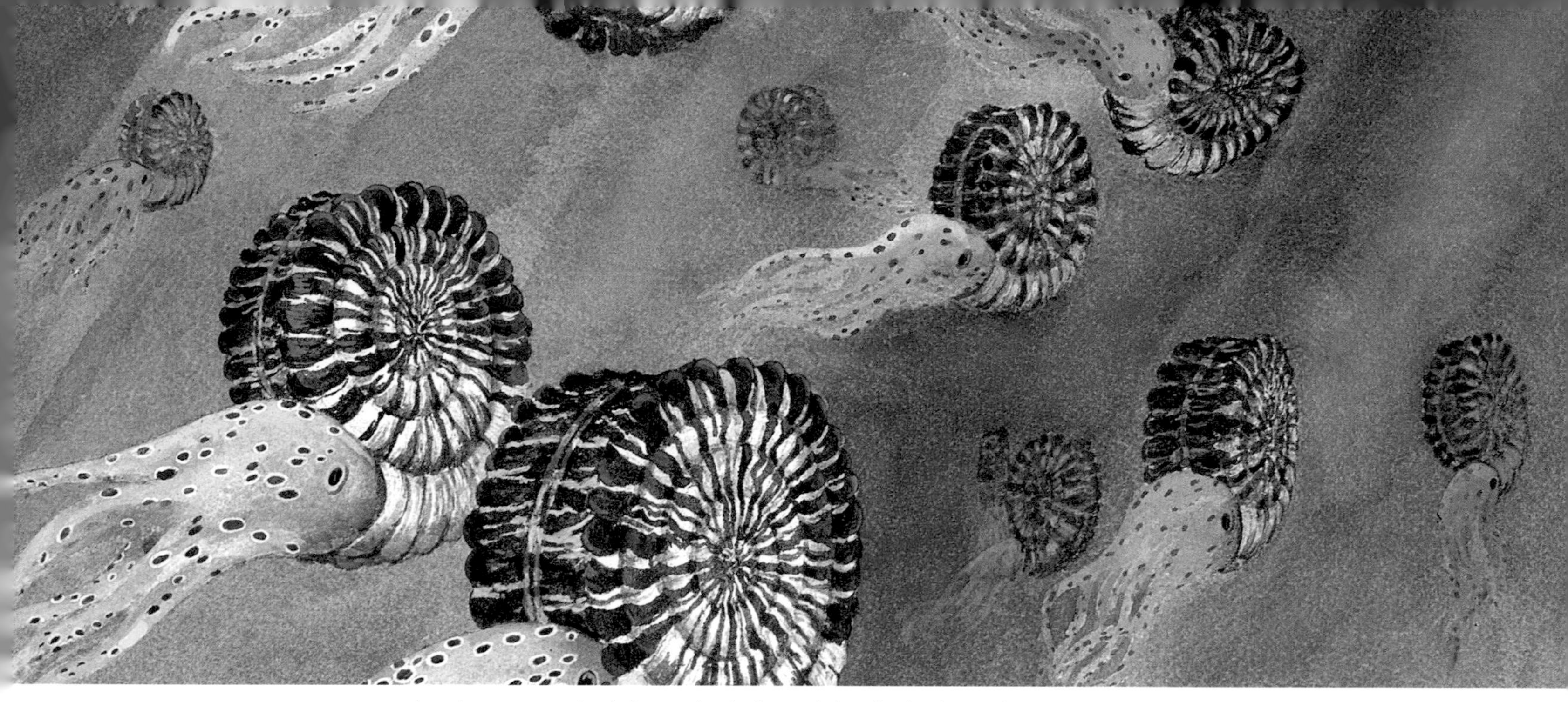

Jurassic seascape: a school of ammonites (extinct cephalopods related to modern octopuses and squids) swimming in a shallow sea about 200 million years ago. Ammonites became extinct at the same time as dinosaurs, 65 million years ago.

CHAPTER 3
Tales of Trails and Ancient Bodies

On the Trail of Ancient Life

One summer day in 1994, brothers Cory and Howard van Allen were beachcombing near the family cottage at Brule, on the Northumberland Strait shore of Nova Scotia. Suddenly, the brothers spotted the remains of tree stumps in rocks exposed by the retreating tide. On closer examination, they found not just stumps, but branches, foliage and fallen trunks. And, just as exciting, they spotted footprints preserved between the trees.

Two Permian *Walchia* tree trunks at Brule, NS. One trunk, to the right, was preserved upright, so that its cross-section is exposed; the other trunk, to the left, has apparently fallen, and is preserved lying parallel to the bedding of the rock, which represents the surface of the ancient forest floor.

Since the van Allen brothers discovered the site, it has been studied in detail, and the results reveal a fascinating story. We now know that the Brule find is the only preserved *Walchia* forest yet discovered. *Walchia*, a type of extinct tree that resembles the modern Norfolk Island pine,

A clam (bivalve) shell, related to the modern quahog, in sandstone of possible Miocene (24-5 million years ago) age, from Georges Bank. The shell is relatively young, as fossils go, and its material (calcium carbonate) remains largely unaltered.

flourished about 285 million years ago, in the Permian. From a comparison with footprints and skeletons found elsewhere, the trackways in the *Walchia* forest appear to be of small, lizard-like animals and large, fearsome, sail-backed creatures. In one example, many trackways made by the same species head off in the same direction. This is the oldest known record of animals moving in herds. By studying the rocks at Brule and those of the same age elsewhere, we know that the *Walchia* forest trees grew in an equatorial monsoon climate with alternating wet and dry seasons.

The preserved tree trunks and trackways found by the van Allen brothers are fossils, examples of the remains of ancient life. As early as about 400 B.C., the Greek scholar Xenophanes realized that many fossils were animal remains. Like a detective, he deduced that fossils found high in the mountains proved that what was now land had once been sea. The discoveries of Xenophanes were ignored in Europe until the Renaissance—more than 2,000 years later—when their real origin was discerned anew by that original "Renaissance Man", Leonardo da Vinci. But even before Leonardo, in the third century A.D., Chinese scholars had correctly identified fossil pine trees, and three hundred years later, they described "stone" fish.

Conrad Gesner (1516-1565), a Swiss naturalist, published *A Book on Fossil Objects, Chiefly Stones and Gems, their Shapes and Appearances*, the first European work that systematically described and illustrated fossils. The word "fossil" originally meant "dug up"; in Gesner's day, this included anything dug out of the earth, even a mineral. But the concept has changed, so that "fossil" now usually refers to the preserved remains or evidence of a living organism. The science of fossils is called "paleontology", and those who study fossils are "paleontologists".

A snail (gastropod) in sandstone, of possible Miocene age from Georges Bank. In this specimen, the calcium carbonate shell has been completely dissolved to form a mold. Before the shell dissolved, the shape of the body cavity had been preserved as a cast composed of finer sediment.

How Fossils are Preserved

To be fossilized, a dead plant or animal must be rapidly buried in sediment or sink into a bog before it decays, is eaten, or erodes. Even then, the chances of soft parts being preserved are remote. Very few fossil hunters will find a complete animal, such as the woolly mammoth, frozen in Siberian soil. A few more will discover whole insects preserved in amber. But most will find only the more durable parts of animals and plants. These include shells of invertebrates (animals without backbones), skeletons of vertebrates (animals with backbones), and wood and pollen grains of plants. Most sediments are deposited underwater in lakes, rivers or seas, and the land is usually an environment of erosion rather than deposition. Thus, most fossils are of aquatic organisms, even though the most famous fossils of all, the dinosaurs, were land creatures.

When the hard parts of animals and plants are buried without any change, we have complete or unaltered preservation. The bones of dinosaurs are commonly preserved in this way. More often, however, the fossil is mineralized to some extent. It becomes "permineralized" when new minerals, usually from ground water, fill empty pore spaces and the original shell, bone or wood stays unaltered. "Replacement" is when the bone or shell material is replaced by minerals. "Petrification" involves both permineralization and replacement. Many plant remains are carbonized (changed to charcoal) so that they appear as black coaly films. Another common type of fossil is a "mold" formed when, in solid rock, the preserved hard parts are later dissolved by ground water. The mold may remain as a hole in the rock or it may be filled by sediment or minerals carried by ground water, forming a "cast". Although soft-bodied animals such as jellyfish and worms are rarely preserved, occasionally we find their impressions made on soft sediment that was soon afterward turned into rock, such as the Cambrian jellyfish-like impressions, about 500 million years old, in Saint John, NB.

Carbonized and compressed *Neuropteris* fern leaves from the late Carboniferous of Cape Breton.

The Microscopic World

Many organisms are too small to be studied by the naked eye. Fossils of these microscopic organisms are called microfossils. Some microfossils, such as diatoms, are the remains of single-celled, plant-like organisms (or "algae") that produce their own food through photosynthesis. Others, such as foraminifera, are single-celled, animal-like organisms (protozoa), actively swimming about and ingesting food. Both algae and protozoa are nowadays referred to as protists, since most are not closely related to true plants and animals. Indeed, some groups, such as dinoflagellates (single-celled organisms that move by means of two whip-like flagella) are both animal-like and plant-like. Other microfossils, such as ostracods, are the relics of tiny, many-celled animals. Still others, like pollen and spores, are the parts of plants. Many microfossil groups have relatives that live as plankton in modern oceans. Plankton are the

A petrified tree stump of late Carboniferous age from Inverness, NS, found weathered out of the strata and stood upright. The tissues of the plant have been infilled with, and replaced by, minerals.

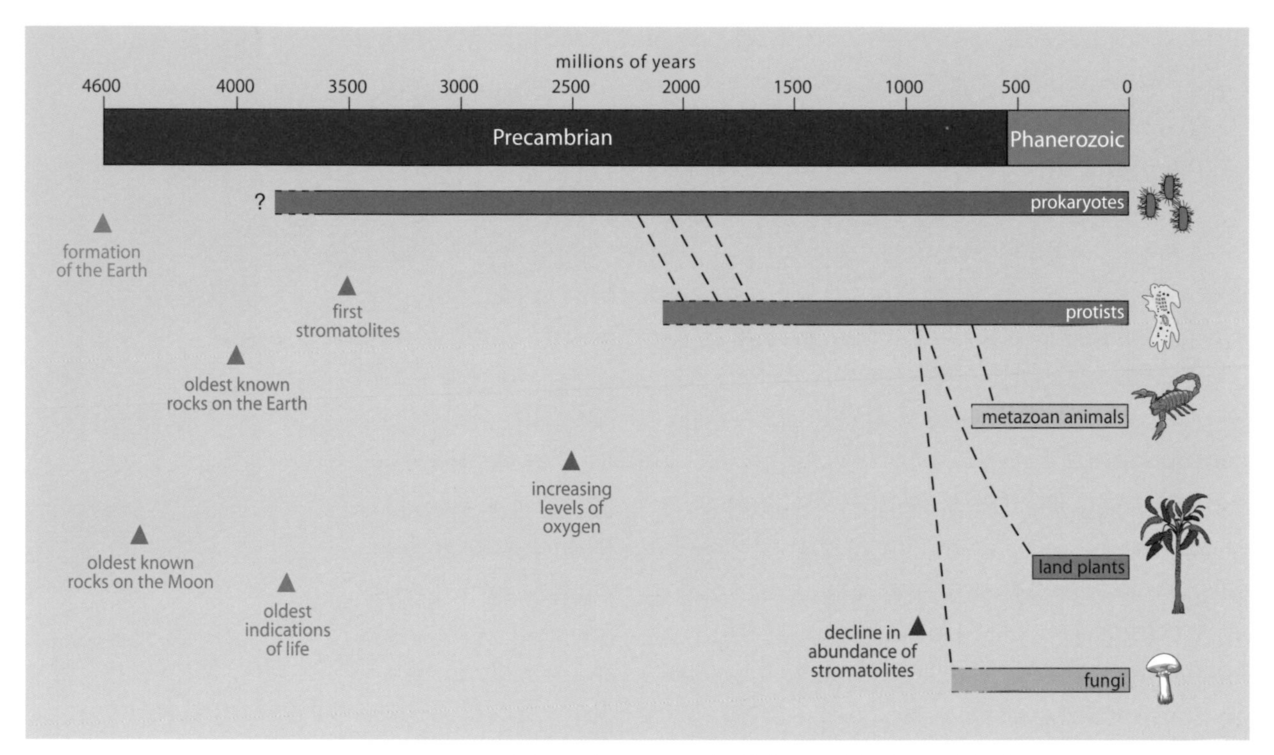

Relationships among major life forms and the timing of significant events.

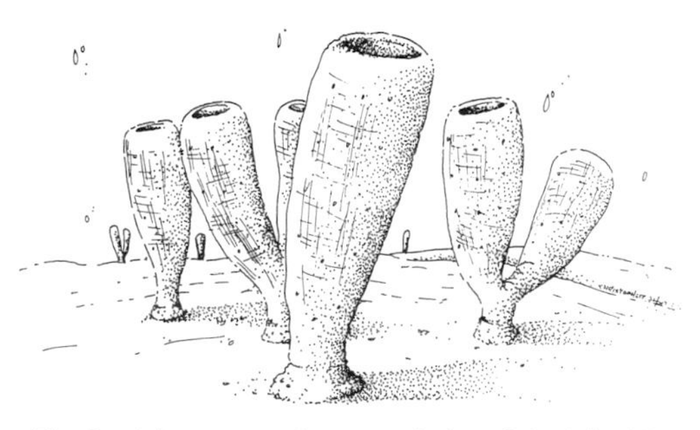

The Cambrian sponge, *Protospongia*, from Saint John, NB. Sponges are among the oldest and simplest invertebrates, each sponge being little more than a colony of well co-ordinated single cells on a framework of chitinous, calcareous or siliceous material. It is this resistant framework that we find preserved as fossils.

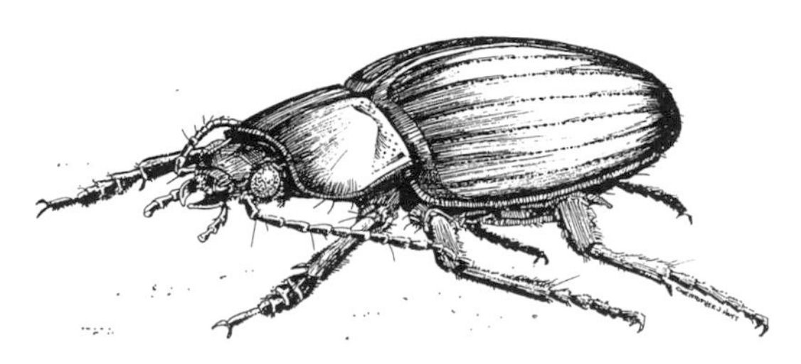

The Holocene insect *Stereoceras*, found at Saint John, NB.

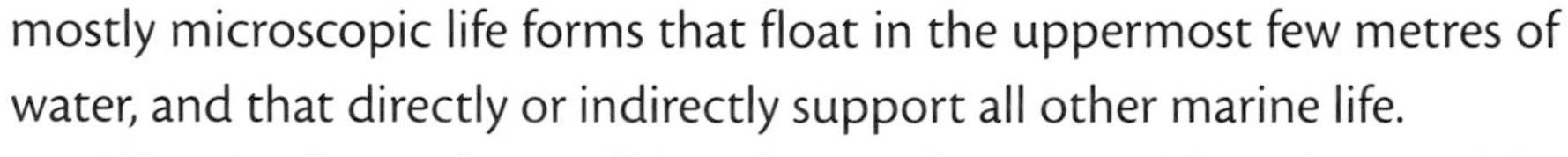

mostly microscopic life forms that float in the uppermost few metres of water, and that directly or indirectly support all other marine life.

Microfossils may be small, but they can have a significant impact. The white cliffs of Dover consist mainly of coccoliths, the protective calcareous (calcium carbonate) plates of a protist. Each small piece of chalk is made up of millions of such plates. The pyramids of Egypt are built of limestone blocks that are mainly the calcareous shells of foraminifera. Some microfossils have shells made from silica, the same material as glass. Microfossils of this type include diatoms that, when concentrated, form the rock diatomite, an example of which is found at Sandy Cove, NS.

Fossil Animals Without Backbones

Our first impression of the animal world is that the vertebrates are dominant. However, in number of species, individuals and life styles, the invertebrates are much more successful. Invertebrate shells can be formed of calcium carbonate, calcium phosphate, or chitin (a tough, organic substance). Shells of calcium carbonate and calcium phosphate are commonly found as fossils, but chitinous shells rarely fossilize. By far the most abundant invertebrates are the arthropods (animals with "jointed legs"). About 75 percent of all modern animals belong to this group and, of these, 75 per-

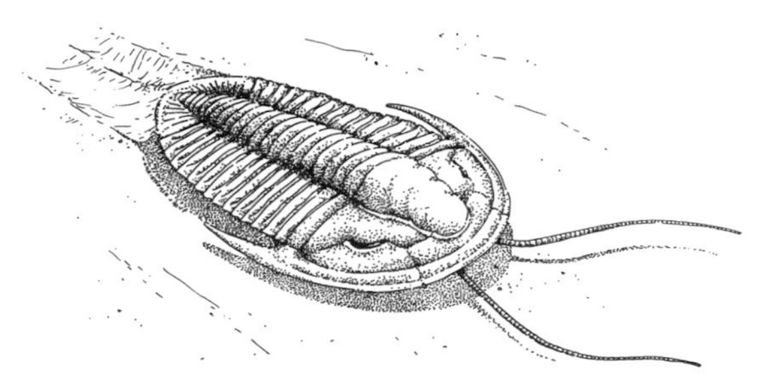

The Cambrian trilobite *Ptychoparia*, leaving behind a trail. Such trails are preserved as fossils and are given the name *Cruziana*.

A Cambrian trilobite burrowing into sediment on the ocean bottom.

cent are insects. Since insects usually have a chitinous outer shell, or exoskeleton, their fossils are rare. One spectacular example is the huge dragonfly-like *Meganeura*, found in the late Carboniferous rocks at Joggins, NS. Smaller insects of similar age, such as *Xenoneura*, have been found at Fern Ledges, in Saint John, NB.

During the Cambrian Period, when hard-shelled animals first became common, the dominant arthropods were exotic marine dwellers such as

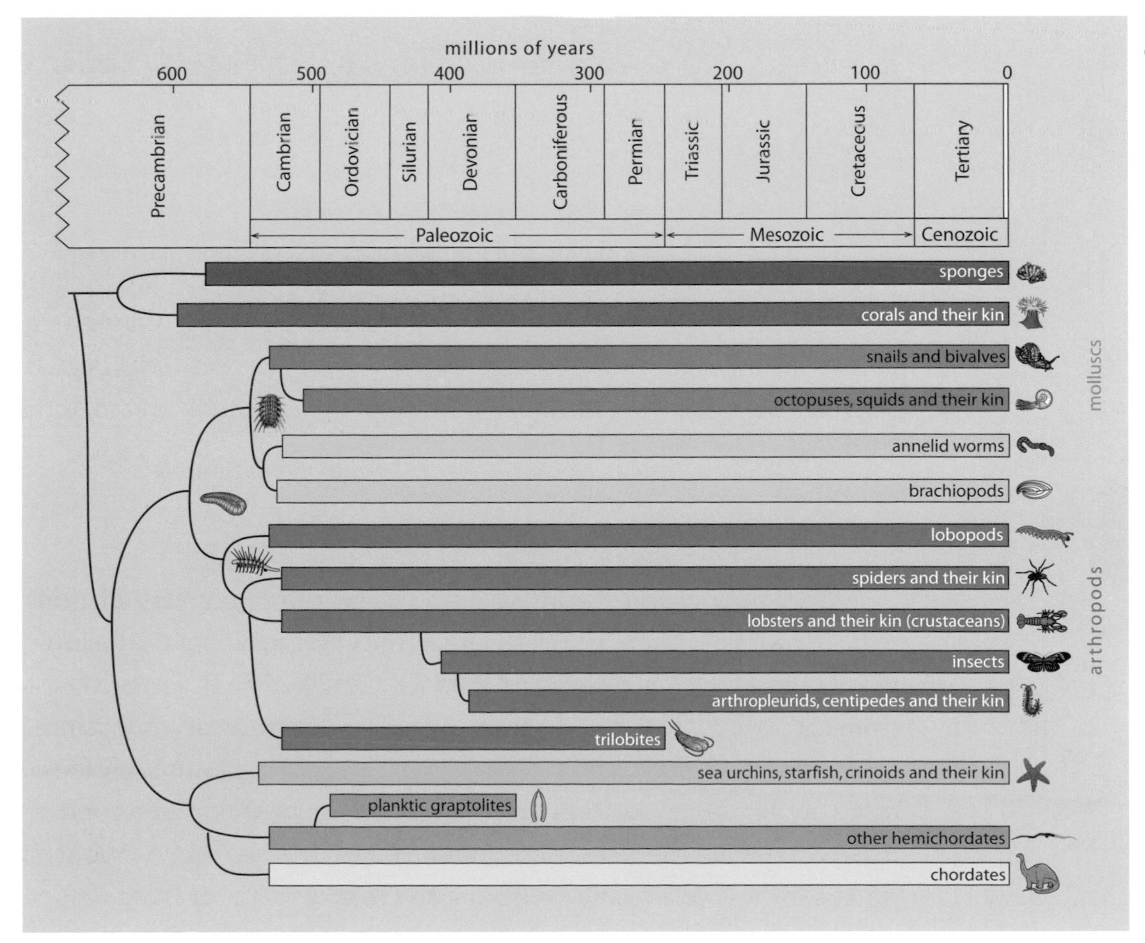

Relationships among animals, with emphasis on the invertebrates.

A Silurian trilobite, *Dalmanites*, from Arisaig, NS.

trilobites (animals with "three lobes"). These invertebrates had a hard, segmented exoskeleton made up partly of calcium carbonate. Because of their segmented bodies, many trilobites could curl up into a ball like modern woodlice. Most trilobites lived on the sea floor in shallow water, grubbing around for food. Some burrowed into the sea floor, and a few were swimmers. There are no trilobites today, because they died off in the greatest of all extinction episodes, at the end of the Permian (about 250 million years ago). Their fossils have been found at several Maritime localities, including Saint John, NB and Arisaig, NS.

Another large group of shelly fossils is the molluscs, which includes snails and slugs (gastropods), clams (bivalves) and the octopus and squid group (cephalopods). Some modern molluscs are the largest and heaviest invertebrates known—there are giant squids about 20 metres long, as well as the giant clam *Tridacna*, which can weigh in at over 200 kilograms. The most adaptable molluscs are the gastropods (animals with a "stomach foot"). Gastropods usually have a single, calcium carbonate shell, which is generally coiled, and they move slowly on their one foot. The ancestors of gastropods first appeared during the Cambrian "explosion", and they are still common today. The earliest known land snails have been found in the late Carboniferous rocks, again at Joggins and Fern Ledges.

A Silurian straight-shelled nautiloid from Arisaig, NS. The chambers of the shell, which were once hollow, are now filled with the mineral calcite.

Bivalves (or clams) are molluscs with two calcium carbonate shells ("valves"). Each valve is generally a mirror image of its partner and is usually of the same size. Any seafood lover who has eaten mussels, clams or scallops is familiar with bivalves. They have a foot, a mouth, and a stomach, but no head. Some are burrowers or borers, some cement or fasten themselves to rocks, and a few swim short distances by clapping their shells together. Most bivalves are marine, but some live in fresh water. They first appeared about 30 million years after the beginning of the Cambrian and are found, for example, at Arisaig, NS (marine forms), Joggins, NS and Chipman, NB (fresh or brackish water forms) and in the sediments of Ice Age beaches.

The most sophisticated group of molluscs is the cephalopods (animals with a "head foot"). Today this group includes squids, octopuses and nautiloids (represented by the modern shelled *Nautilus*). Cephalopods have a unique swimming style: they move by jet propulsion, ejecting a powerful stream of water. And those with chambered shells can sink or rise by using their chambers as floats, filling or emptying them with water to change weight. All cephalopods are marine, and most are carnivores. They appeared about 500 million years ago, a little later than the first bivalves.

Among the earliest known cephalopods in the Maritimes are the straight-shelled nautiloids, found in Silurian rocks at Arisaig, NS.

Perhaps the best-known fossil cephalopods are ammonites (animals shaped like the horn of the Egyptian god, Amon). When ammonites were first found, hundreds of years ago, they were thought to be the remains of sea serpents. Most ammonites look like coiled rope, whereas others have straight or curved shells; all consist of numerous chambers, with the animal living in the outermost chamber. Ammonites became extinct at the same time as the dinosaurs. They are occasionally found in rocks from oil exploration wells drilled on the Scotian Shelf.

Brachiopods (animals with an "arm-foot") appeared at almost the same time as trilobites. Like bivalves, they have two shells, and these are made of calcium carbonate or calcium phosphate. Brachiopods are distinguished from most bivalves in that the two valves of an individual are not mirror images of one another. In brachiopods other than primitive forms, the two shells are hinged. Brachiopods commonly fasten themselves to the sea floor by means of a stalk, hence the name "arm-foot". Although brachiopods first appeared during the Cambrian, they are still living today—in the Bay of Fundy, for example. But they are now rare, in contrast to their abundance in Paleozoic seas. Maritime brachiopods are found in early Cambrian rocks near Saint John, NB; in Silurian rocks at Arisaig, NS; and in many marine Carboniferous rocks.

Some of the simplest but best known invertebrates are corals. These animals can live in colonies or as solitary individuals. Those most commonly preserved as fossils have a calcium carbonate skeleton. Corals are famous for building reefs and are thus an important rock-building group. Many limestones were formed from their skeletons. Although the colonial corals that form reefs are indicators of warmer water, solitary corals live today in cooler temperatures, even as low as 1°C. In Nova Scotia, fossil corals are found in Carboniferous rocks of the Windsor area. In New Brunswick, they are present in Silurian and Devonian strata along Chaleur Bay.

Another important group of invertebrates are the echinoderms, such as the sea urchins (echinoids), starfish (asteroids) and sea lilies (crinoids and eocrinoids), most of which live on the sea floor. They are the only invertebrates with a five-fold radial symmetry. Echinoderm means animal with a "spiny skin". This is a good description for a sea urchin, which has a covering of imposing, sharp spines. Sea urchins have a multitude of delicate, tentacle-like feet that serve a variety of purposes, including walking, feeding and breathing. Sea urchins are found in Quaternary marine clays in

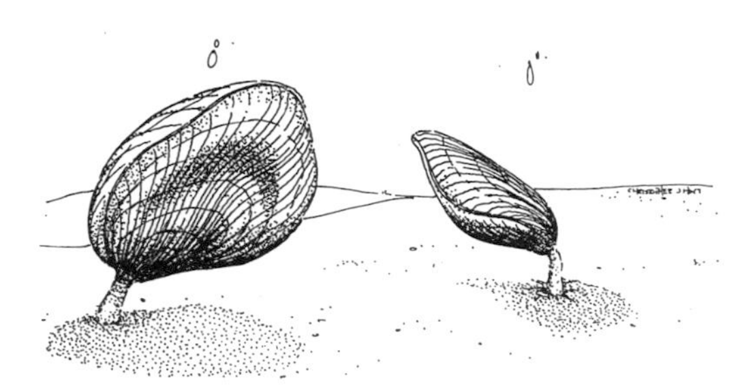

The Silurian-Devonian brachiopod *Atrypa*.

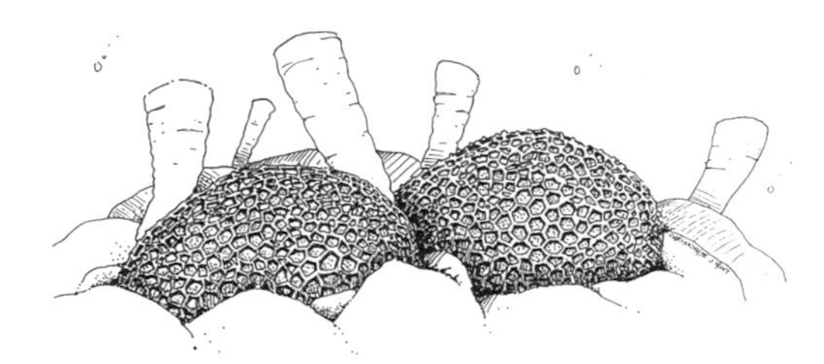

Two modes of life among corals, solitary forms (the higher horn-like structures) and the mound-like colonial *Favosites*, in early Devonian rocks at Dalhousie, NB.

A large solitary coral of early Carboniferous age found at Newport Landing, near Avondale, NS. In life, the coral had a ring of tentacles extending upwards, like those of a sea anemone.

The sea urchin (echinoid) *Strongylocentrotus*, found in Quaternary sediments at Point Lepreau, near Maces Bay, NB.

Saint John, NB. Crinoids and eocrinoids are stalked animals, each with a stem attached to the sea floor. The stem is composed of discs or plates, and these are most often found separately as fossils. At the free end of the stem is a cup (calyx) in which the animal lives. Both crinoids and echinoids first appeared about 475 million years ago. Crinoids are common in Silurian rocks near Arisaig, NS and near Dalhousie, NB. Eocrinoids are found in Cambrian rocks near Hanford Brook, NB.

Our closest invertebrate relatives are those animals that have a spinal cord protected by a stiff non-bony rod (called a "notochord") rather than a backbone. Such animals include the amphioxus *Branchiostoma*, which closely resembles the fossil *Pikaia* from the Cambrian Period. All animals with a spinal cord, regardless of whether or not it has a protective backbone, are called "chordates", a more fundamental biological category than that of vertebrates.

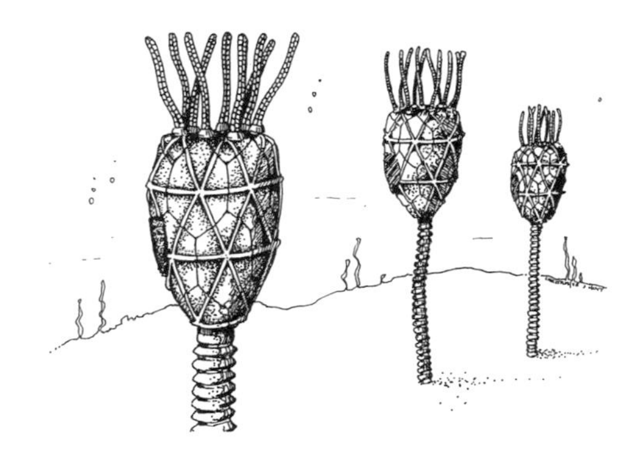

An early echinoderm, the eocrinoid *Eocystites*, from Saint John, NB.

The graptolites are a group of fossils that may represent animals distantly related to the vertebrates. They are similar to a group of living organisms, the pterobranchs, that grow in ways similar to chordates but have neither a backbone nor a notochord. Graptolites and pterobranchs are therefore called "hemichordates" ("half-chordates"). The name graptolite, which means "written stones", reflects their appearance as saw-toothed black lines on rock. Graptolites lived in colonies made up of rows of cups strung together, thus creating the saw-toothed pattern. They were either anchored to the sea floor (middle Cambrian to Carboniferous) or floated in the water as planktonic colonies (Ordovician to early Devonian). It is the planktonic forms that are most commonly found as fossils. An example is *Rhabdinopora*, which occurs in early Ordovician black shales in Saint John, NB.

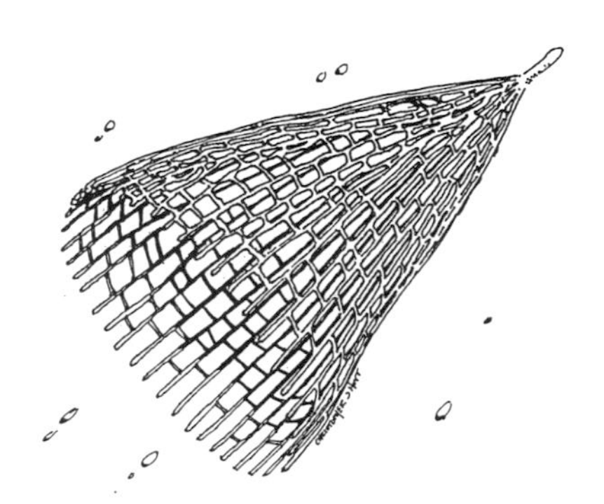

The Ordovician graptolite *Rhabdinopora*, found at Saint John, NB and near Kentville, NS.

Animals With a Backbone

Life evolved in water; the land was too barren and hostile until plants became established there some 450 million years ago. Thus the oldest vertebrates are fish. The earliest fish are found in Cambrian rocks about 520 million years old. They had a cartilaginous rather than a bony skeleton, and lacked jaws, simply having a hole for a mouth, like the present-day lamprey. By about 400 million years ago, in the Devonian, jawed fish were abundant. Toward the end of the Devonian, the first vertebrates, the amphibians came onto land. Unlike fish, these animals could survive out of water, but had to return to water to lay their eggs, just as their modern descendants, the frogs, still do. However, the first amphibians did not resemble frogs, and were more like large salamanders. Some were about a metre long with flat heads, formidable jaws and lots of teeth.

If there were amphibians roaming the land 380 million years ago, when did the first reptiles appear—and what is so unique about them? Reptiles have a big advantage over amphibians: they evolved a hard-shelled egg to protect the developing embryo. This "survival capsule" meant that reptiles were not tied to the water, and could roam free on vast tracts of unclaimed Paleozoic land. The Maritimes has played an important role in this story. *Hylonomus*, ("forest mouse"), found inside 310 million-year-old tree trunks in the cliffs at Joggins, NS, was for a long time the oldest known reptile. Reptiles found in later Paleozoic rocks of the Maritimes include the possibly sail-backed mammal-like reptile *Bathygnathus*, from Prince Edward Island.

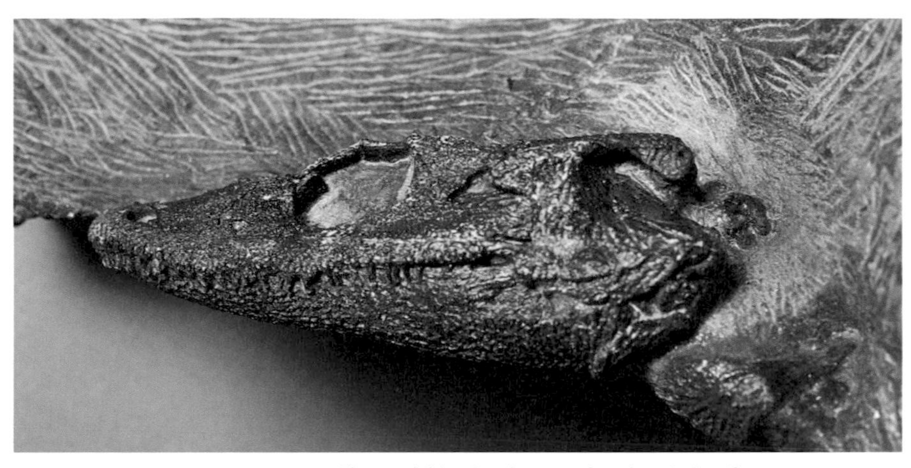

The amphibian *Dendrerpeton* from late Carboniferous rocks of Joggins, NS, showing details of the skull.

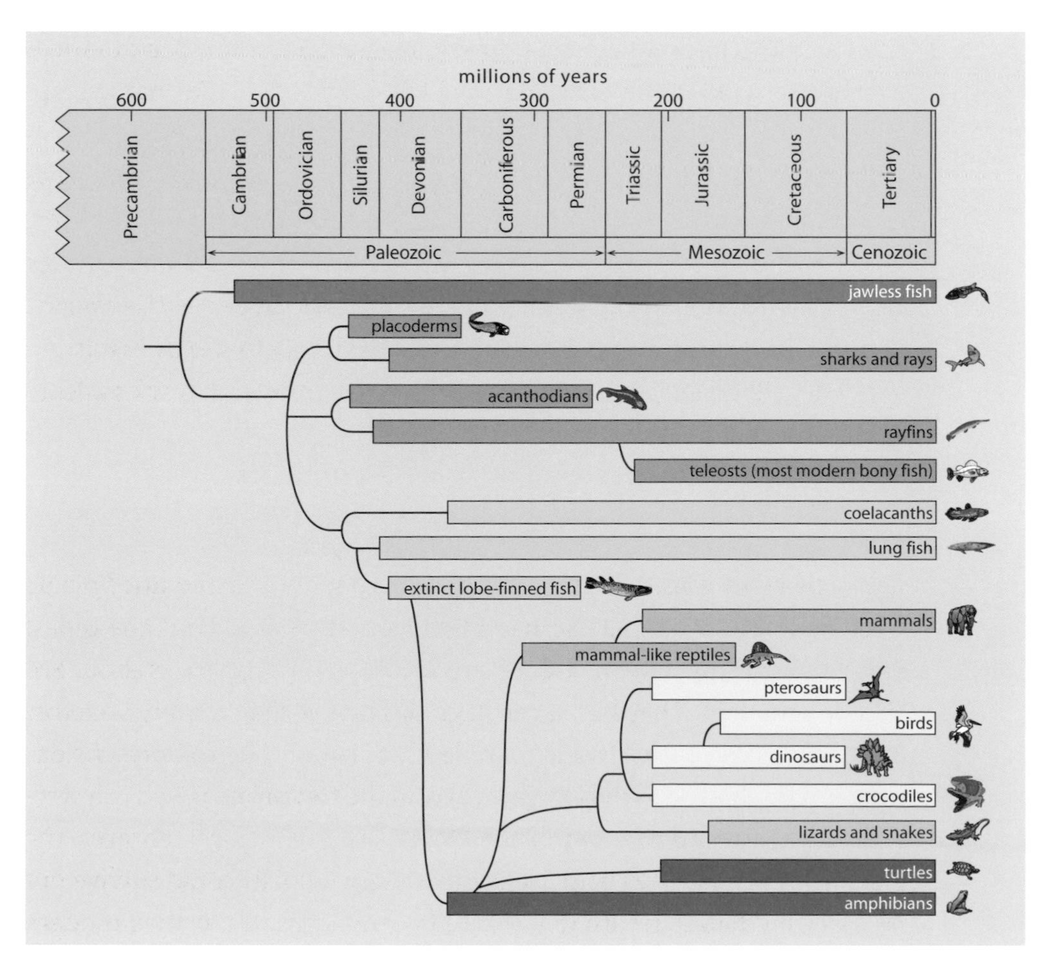

Relationships among vertebrates.

An early reptile gazes out of a clubmoss tree stump as a forest fire approaches.

Lesothosaurus, an early Jurassic dinosaur. Dinosaurs were reptiles, as shown by their scaly skin and shelly eggs.

At the end of the Permian, which is also the end of the Paleozoic, a major catastrophe killed most life, including many reptiles. The reptiles that survived went through a great diversification in the Mesozoic era that followed, as the supercontinent Pangea broke up. In fact, these descendants were so successful that the Mesozoic has been called the "Age of Reptiles". Mesozoic reptiles included the flying pterosaurs and the sea-going ichthyosaurs and plesiosaurs. However, the most famous of all Mesozoic reptiles are the dinosaurs, some of which were the largest land creatures known. Dinosaurs were unusual reptiles because of the way they walked: they had an erect posture, in contrast to the sprawling gait of most other reptiles.

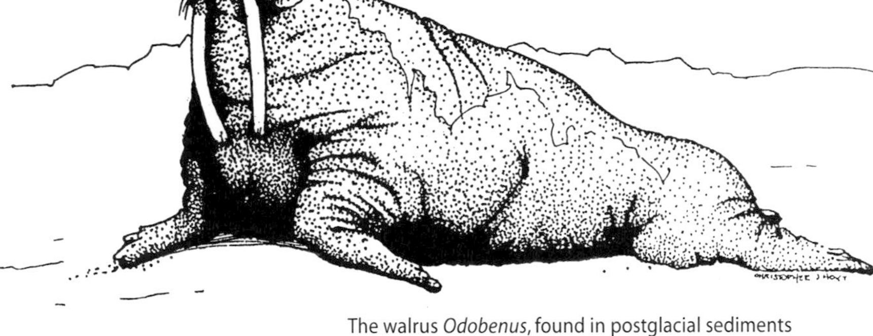
The walrus *Odobenus*, found in postglacial sediments of the Bay of Fundy and the Scotian Shelf.

In the Maritimes we have evidence for only part of the dinosaur story—but it is a critical part. On the shores of the Bay of Fundy, bones of some of the oldest known dinosaurs have been found. These are about 220 million years old. Also on the Fundy shore, at Wasson Bluff near Parrsboro, NS, are the remains of 200-million-year-old mammal-like reptiles, as well as crocodile-like creatures, and *Clevosaurus*, a small lizard-like reptile of a group that is represented today only by the tuatara of New Zealand. At the end of the Mesozoic, dinosaurs and some other groups died out over a relatively short period of geological time. But recent spectacular fossil finds from China seem to confirm that birds are the descendants of dinosaurs. So, in a sense, dinosaurs are still among us.

During the Mesozoic, mammals were a small, relatively inconspicuous group. Mammals differ from reptiles in having hair or fur, in usually giving birth to live offspring and in providing milk for their young. In the Cenozoic, mammals quickly filled the ecological niches left after the extinction of many of the Mesozoic reptile groups. By the Ice Age (the last two million years), the mammals found in the Maritimes would have been more or less familiar to our modern eyes. The mastodons found at Hillsborough, NB, and Milford, NS, are clearly members of the elephant family. And the fossil walruses found on the Fundy shoreline of New Brunswick and on the Scotian Shelf are the same as modern walruses, even though these animals live much farther north today.

Fossil "Behaviour"

Reptilian footprints and tail drag mark from latest Carboniferous to Permian rocks at Point Prim, PEI.

Burrows, tracks, trails and footprints provide clues about how an animal moved or lived—in other words, about its behaviour. Equivalent structures, called "trace fossils" are found in sedimentary rocks. Footprints are particularly useful for providing clues about an animal. By measuring the footprints and the trackway as a whole, and by noting the direction, paleontologists can tell how big the animal was, whether it was a reptile or amphibian, if it was running, walking, or even limping, and its speed. Where the footprints of more than one animal are preserved, it may be possible to find out how animals interacted with

Sketch of an amphibian and its footprints.

each other. For example, we know that some dinosaurs lived in herds because of the numerous trackways found together. And, as mentioned earlier, the trackways at Brule, NS, tell a story of a similar herding lifestyle almost 100 million years before the first dinosaurs.

In the Maritimes there are some spectacular footprints of prehistoric animals. These include tracks of early four-legged land vertebrates in Carboniferous rocks near Hantsport, NS, and near Sackville, NB, as well as tracks of early dinosaurs in rocks on the Fundy shore near Parrsboro, NS. On Prince Edward Island, trackways and tail drag marks have been found in the latest Carboniferous-Permian rocks in the Point Prim area.

Many fascinating facts about ancient life come from studies of fossil droppings or dung. If the dung is from a herbivore, then plant remains, including spores and pollen, will show what plants the creature ate. Fossil dung found with the Hillsborough mastodon were used to investigate that animal's diet and surroundings. For a carnivore, bones in the dung may give clues about the animals eaten by the long-gone dung-maker. The scientific word for fossil dung is "coprolite".

Green Revolutions

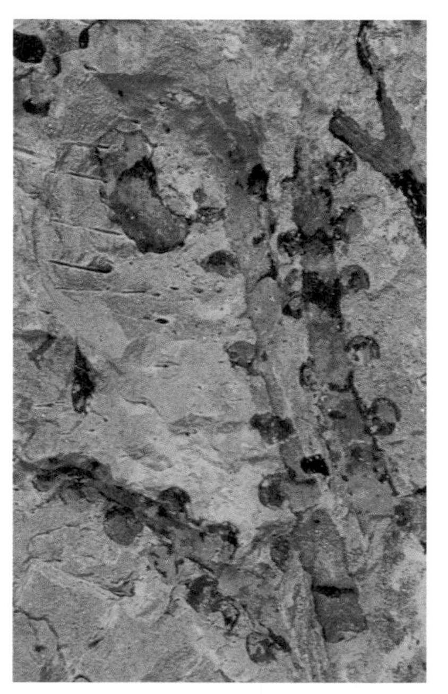

Stalks and attached sporangia of the Devonian plant *Oricilla*, from Dalhousie Junction, NB. This plant is preserved as carbonized material (black) and iron oxides (orange).

Hundreds of millions of years before true plants moved onto land, plant-like protists (algae) were growing and multiplying in the sea. They were the main source of the oxygen in the atmosphere, and they became the main food source for many animals. Surviving in air is more difficult for plants and animals than surviving in water. In air, there has to be protection against water loss and intense ultraviolet light, as well as an ability to cope with greater temperature extremes. If an organism is to grow tall, it must have a support system for withstanding wind and gravity, and an internal method for distrib-

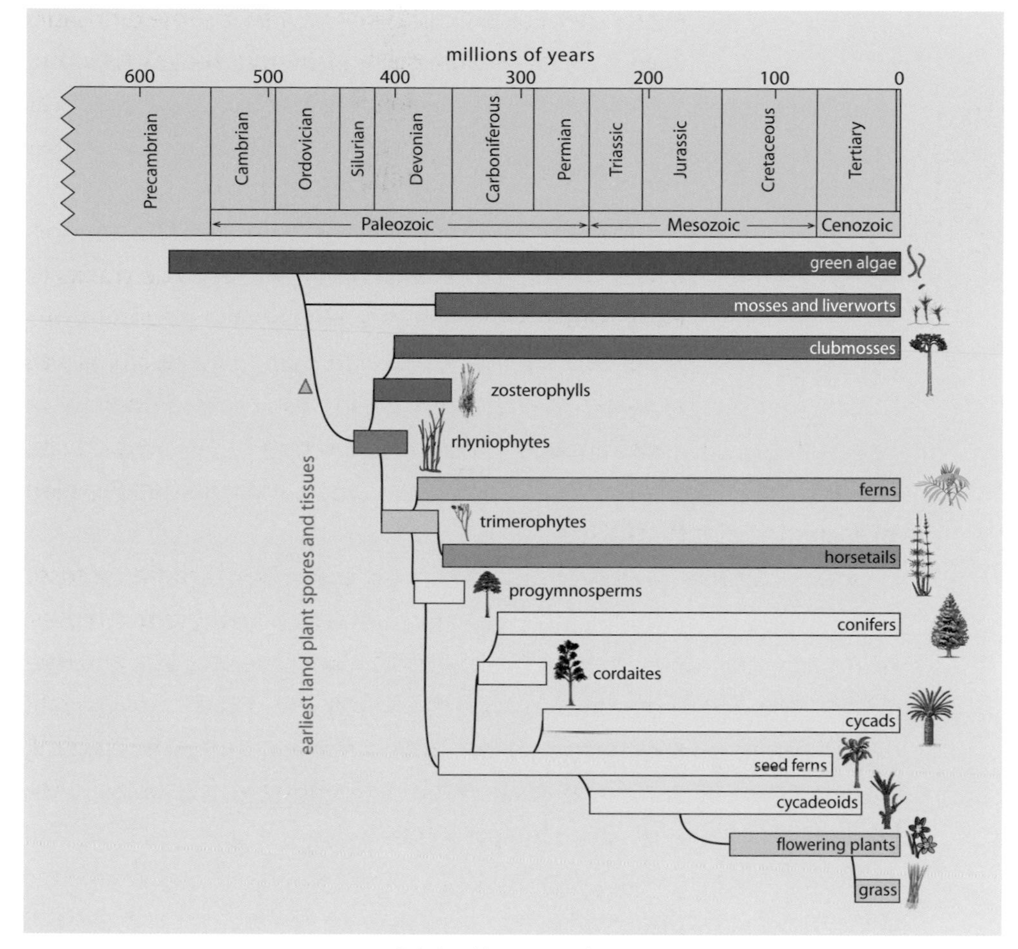

Relationships among plants.

uting water and nutrients. The plants that best solved the problems of surviving on land are the vascular plants, so called because of the development of a vascular system. This system of tiny tubes for conducting water and food throughout a plant's body is analogous to the veins and arteries of animals, which is also called a vascular system.

Plants, other than perhaps lichens, made the impressive leap to land about 450 million years ago, in the Ordovician. The earliest evidence for land plants comes from spores and tissue fragments. These may have come from mosses and liverworts (or "bryophytes"). Bryophytes didn't develop a vascular system, so they stayed small and, to this day, must live in moist environments. Vascular plants were present by the middle of the Silurian, about 430 million years ago. The earliest vascular plants are known as rhyniophytes, after the Scottish locality of Rhynie, where they were discovered. Rhyniophytes consisted of simple stems with knob-like sporangia (spore-bearing organs) at the tips. They did not have leaves. Remains of early plants are abundant in Devonian rocks at Dalhousie Junction, NB.

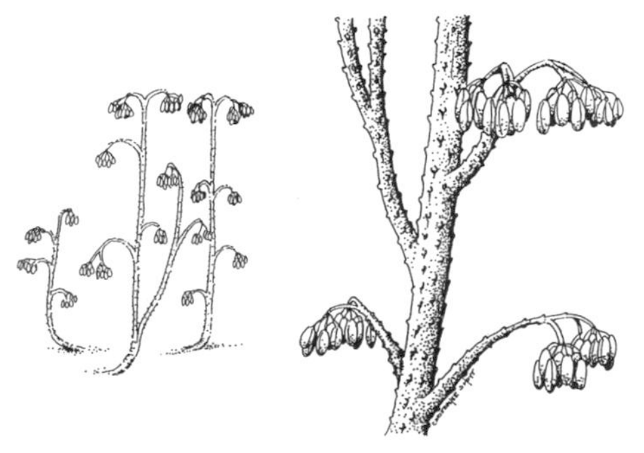

Psilophyton, a Devonian plant from Dalhousie Junction, NB.

Early vascular plants soon split into two main lineages. One of these led to a group of plants known as clubmosses, represented today by the moss-like plants *Lycopodium* and *Selaginella.* The other ("mainstream") lineage led to all other vascular plants, including horsetails and hemlocks, ferns and flowers.

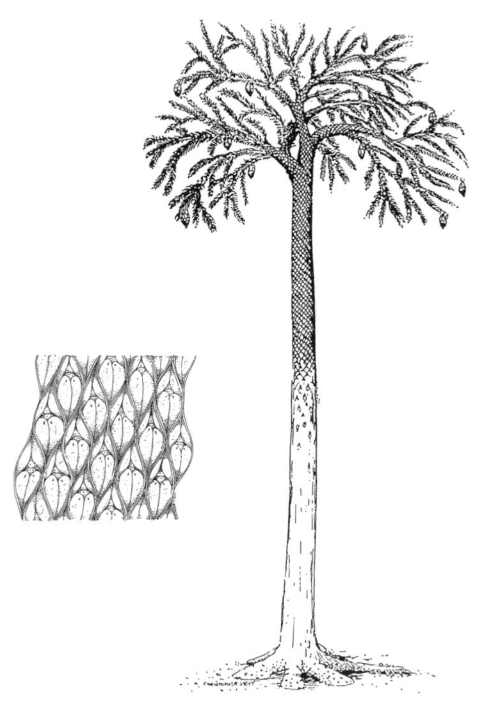

Lepidodendron, a giant Carboniferous clubmoss, which is common, for example, in the cliffs at Sydney Mines, NS. The inset shows leaf scars on the bark.

Although inconspicuous today, clubmosses were prominent in Carboniferous forests. Indeed, fossil tree stumps in cliffs at such places as Joggins, NS, and Cape Enrage, NB, are giant clubmosses. Clubmosses of the Carboniferous, such as *Lepidodendron* and *Sigillaria*, grew into great trees up to 50 metres tall. However, their trunks were not like those of other plant groups. Rather than having a core of solid wood surrounded by softer bark, clubmoss trees were supported by a thick bark. Their trunks had very little xylem—the tissue that conducts water from roots and that forms the wood in most modern trees. And there was no phloem—the tissue that conveys food materials around most plants, chiefly from the leaves, where food is manufactured. Perhaps the scaly leaves on clubmoss stems, or the stems themselves, may have produced food to sustain stem tissues. If so, clubmoss tree trunks would have been green rather than brown. Because the outer parts of clubmoss tree stems were tougher than the insides, their fossil tree trunks are usually preserved as cylinders of preserved bark. The inside of the live tree consisted mostly of softer pithy tissues. After the tree died, these tissues decayed and were replaced by sediment. Fossil roots of clubmosses, known as *Stigmaria*, are also commonly preserved.

Horsetails, recognizable by their jointed stems, were another type of plant prominent in Coal Age forests. The most common example is *Calamites*. Like clubmosses, horsetails were much more abundant than they are today, represented in modern floras by *Equisetum*, a widespread, small, non-woody plant that grows in moist places and poor soils. Ferns also grew as trees and shrubs in Coal Age forests and throughout the later

Segment of the clubmoss tree *Sigillaria*. The trunk, from late Carboniferous rocks of Glace Bay, NS, is preserved as a mudstone-filled cast with carbonized (coal) bark on the outside.

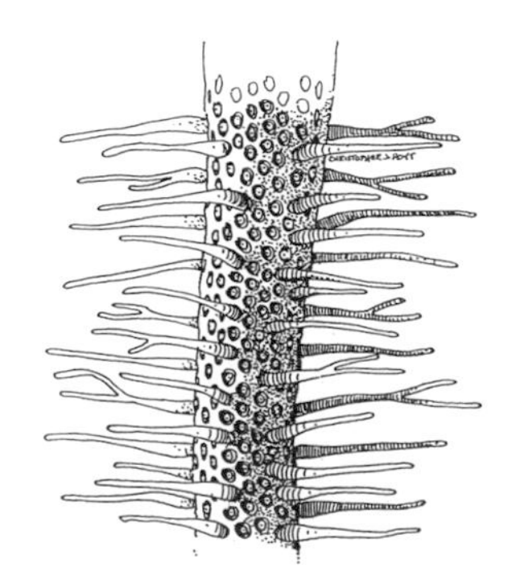

The late Carboniferous root *Stigmaria*, with rootlets. (Fossil plants are rarely found whole. Parts of plants, such as trunks, roots, leaves and cones, are usually found scattered, so only detective work can piece together the original plant–and even then we are left with many conundrums. Because of this, different parts of fossil plants are given different names. For example, the name *Stigmaria* is given to the roots of both *Lepidodendron* and *Sigillaria*, which are indistinguishable.

Mesozoic. They are still common today, but survive mostly as shrubs rather than trees; modern tree ferns are restricted to the tropics.

Clubmosses, horsetails and ferns reproduce by spores, generally microscopic bodies that are dispersed by wind or water and that require moisture for germination. A major step in the evolution of plants was the appearance of seeds. The precursors of seeds were found in plants that developed spores of two sizes: megaspores and microspores. The megaspores evolved into seeds, and the microspores evolved into pollen grains. Seeds remain protected within the plant, so moisture was no longer needed for fertilization. And pollen grains became tougher, which made them more resistant to drying out, and smaller, which made for easy dispersal.

A late Carboniferous tree fern related to *Medullosa*.

Early seed plants included *Archaeopteris* from the late Devonian and early Carboniferous. In Carboniferous coal forests there were seed ferns, an extinct group of plants that looked like ferns but had seeds, and *Cordaites*. *Cordaites* was a seed plant with strap-like leaves and seeds strung loosely along a fertile branch. In conifers (literally, cone-bearing plants), the fertile branch, or cone, is very short and has scales to protect the seeds. (Cones, however, were not new to conifers—clubmosses also had cones.) Seed plants were common in the Mesozoic and included cycads and cycadeoids, plants consisting generally of a stump-like stem with a circle of palm-like leaves on top.

Cordaitanthus, a seed-bearing branch of the gymnosperm tree *Cordaites*. Late Carboniferous, Westville, NS. (Gymnosperms are essentially seed-bearing plants other than angiosperms.)

About 140 million years ago, one of the most important events in the history of our planet occurred: the appearance of the angiosperms (flowering plants). Much of the evidence for their early history comes from pollen grains found in rocks such as those on the Scotian Shelf. Today, flowering plants are so successful that there are about 250,000 species, compared to less than 600 for conifers (though conifers still dominate vast areas in today's northern forests). A major reason for the success of angiosperms is that many are insect-pollinated. But grass, one of the most successful groups of angiosperms today, is mainly wind-pollinated. Grass is such a familiar part of the modern scene, and looks so unpretentious, that we feel intuitively that it must be ancient. For once, though, intuition fails us, as grass has only been around for the past 25 million years. Because the dinosaurs died out 65 million years ago, one thing is certain: the "terrible lizards" didn't eat grass.

Evolution

Before the middle of the nineteenth century, most scholars believed in the fixity of species. It was thus thought that each species was specially created by God, did not change and could not give rise to other species. One of the first to challenge this idea was the French biologist Jean Baptiste Lamarck (1744-1829), who concluded that there was a natural sequence for all living creatures. He believed that external conditions caused changes in an animal. For example, a giraffe grew a longer neck to reach higher branches, and succeeding generations of giraffes inherited this helpful acquired characteristic. Lamarck's theory was largely ignored because of the immense influence of Georges Cuvier (1769-1832), called in his day the "dictator of biology". Cuvier was a firm proponent of the fixity of species. Although he realized that changes in fossil assemblages over time proved that some species were now extinct, he believed that this was caused by a series of catastrophes.

By mid-Victorian times, the fossil record and geological time were better understood. Plant and animal breeding was becoming a serious science, demonstrating that selection (in this case artificial) within a species could cause great changes in appearance, as seen in the bewildering variety of domestic dogs. The evidence was becoming so convincing that even some of the more scripturally-oriented scientists began to realize that species were not inflexibly fixed. One of these scientists was Richard Owen (1804-1892), the austere superintendent (today we would say curator) of natural history at the British Museum, and creator of the word "dinosaur". He accepted that, although God had created "archetypes" for each major group (the reptiles, for example), within each group a degree of evolution was possible. However, in spite of widespread acceptance of the general idea of evolution, it was not clear how the process worked.

Revolutionary events in science are rare. Examples are the publication of *Principia Mathematica*, in which Isaac Newton forever changed the science of moving bodies, as well as Einstein's special theory of relativity, and the plate-tectonic revolution in geology. On an equal footing is Charles Darwin's *Origin of Species*, published in 1859, a book that contained the first logical explanation of how evolution could occur through the process of natural selection. Natural selection is the concept that, from a variable population of organisms, those with characteristics most suited to their environment will produce the most offspring. Organisms best adapted to their environment will thus become more abundant. If the environment changes, the population will become modified in a different direction.

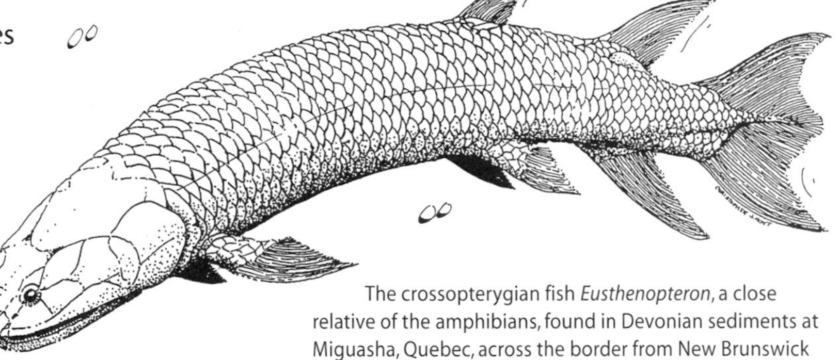

The crossopterygian fish *Eusthenopteron*, a close relative of the amphibians, found in Devonian sediments at Miguasha, Quebec, across the border from New Brunswick

The seed of Darwin's theory of natural selection was born during his five-year cruise on HMS *Beagle* (1831-1836). Darwin, the naturalist on board, was fascinated by the animals living on the Galapagos Islands. Unique species of finches, identified by beak shape and hence food type, lived on individual islands. But all had clearly evolved from common ancestors that had arrived from the mainland of South America, not so long ago. Recently, in a clear demonstration of natural selection, researchers studying "Darwin's finches" have noticed how beak shape can change in response to natural changes in climate and food source. Add vast passages of time to the power of natural selection, and it is easy to see how life could have evolved into its amazing diversity.

All organisms have the same basic chemistry, showing that they are ultimately related. Life processes in everything from bacteria to belugas, trees and tyrannosaurs are based on compounds such as nucleic and amino acids. All four-legged creatures (or "tetrapods", with human arms and birds' wings considered honorary "legs") must have had a common ancestor because the bone arrangement in the limbs is so similar. The existence of fossil fish with this same basic "tetrapod" bone arrangement in their pelvic and pectoral fins suggests that the common ancestor was a fish. Pelvic and pectoral fins must have evolved into back and front limbs, respectively.

Why and how did fish evolve into amphibians? It used to be thought that, as smaller bodies of water dried up during the Devonian Period, fish fins developed into amphibian legs when animals tried to migrate to other water bodies. So animals with the ability to move across dry land were

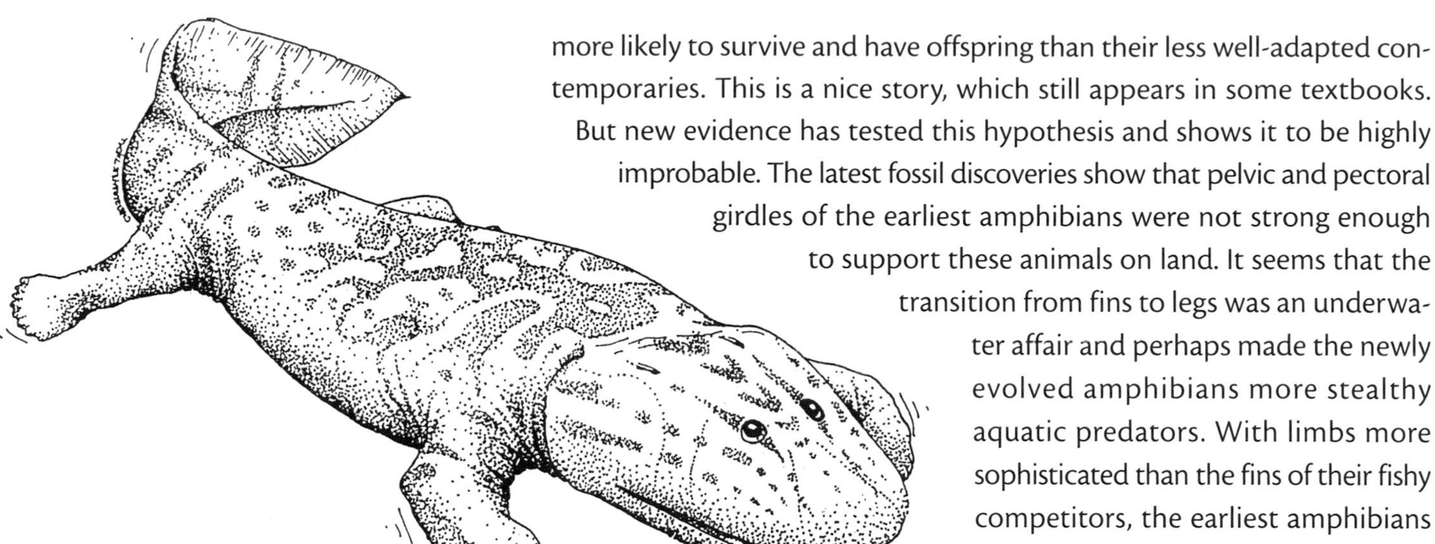
The Devonian amphibian *Acanthostega*.

more likely to survive and have offspring than their less well-adapted contemporaries. This is a nice story, which still appears in some textbooks. But new evidence has tested this hypothesis and shows it to be highly improbable. The latest fossil discoveries show that pelvic and pectoral girdles of the earliest amphibians were not strong enough to support these animals on land. It seems that the transition from fins to legs was an underwater affair and perhaps made the newly evolved amphibians more stealthy aquatic predators. With limbs more sophisticated than the fins of their fishy competitors, the earliest amphibians might have been able to cling to underwater vegetation without an attention-grabbing flapping of fins. Then they could silently move in on their prey.

This fins-to-limbs story illustrates several points. It shows the difference between facts (the fossil bones themselves) and hypotheses (ideas about what caused the changes that we see in the bones). It shows how hypotheses can be modified as new facts come to light. Such informed revisions of scientific ideas are healthy sign of progress in our knowledge. The story also highlights the opportunistic nature of evolution: tetrapod limbs, it appears, did not evolve initially for walking on land. But what a change they made. The evolution of life is full of such opportunism.

Since Darwin, there has been much progress in broadening our understanding of evolution. The most notable advance has been the development of genetics. Gregor Mendel (1822-1884), an Austrian monk, carried out a classic series of experiments with garden peas. He showed that characteristics such as colour and shape were passed on to future generations by small particles that we now call "genes". The discovery of the structure of DNA has further enhanced our understanding of heredity and the evolutionary path. Despite these major advances, Darwin's ideas remain the basis of our understanding of evolution and in general have been overwhelmingly supported by thousands of experiments and observations.

Extinction!

Of all the aspects of evolution, it is extinction that has most captivated the public imagination. We owe much of our knowledge on the subject to Georges Cuvier, who, about fifty years before Darwin's time, proved that extinction is part of the history of life. He based this on a study of living and fossil elephants. Fossils show that, during much of Earth's history, the rate of extinction has not changed much. However, at certain times, there have been "mass extinctions" when this rate has accelerated dramatically. The greatest mass extinction event was at the end of the Permian, about 250 million years ago. At that time, about 95 percent of marine invertebrate species and about 75 percent of tetrapod species died out. Another well-known extinction was the demise of the dinosaurs (along with many other animals), about 65 million years ago, at the end of the Cretaceous. Such fatalities are catastrophic in the short term but open up new ecological niches—the extinction of the dinosaurs, for example, provided an opportunity for mammals to diversify.

Several hypotheses have been put forward to explain mass extinctions. These include reversal of the Earth's magnetic field, comets or asteroids colliding with the Earth, supernovae (the dramatic explosion of dying stars), and changes in the Earth's atmosphere—such as an increase in the amount of carbon dioxide caused by massive volcanic eruptions. Some scientists now believe that one or more of these factors is responsible for mass extinction events. For example, many consider that the effects of a comet or asteroid impact were responsible for the demise of the dinosaurs. However, other scientists, especially paleontologists, are more cautious about linking single catastrophic events to major extinction episodes. In Chapters 6 and 7, we discuss the extinction events at the end of the Permian and at the end of the Cretaceous, respectively.

Today we are in the midst of another major extinction episode, due mainly to *Homo sapiens*. Our success has come at the expense of many other animals, such as the woolly mammoth, the dodo, the passenger pigeon and countless unknown creatures of tropical rain forests. We should remember the sobering lesson of the dinosaurs, who were so successful for about 150 million years. Nothing is forever!

The early dinosaur *Coelophysis* appears to have died out in a major extinction event at the end of the Triassic; this was about 135 million years before the end-Cretaceous event that killed off the last dinosaurs. The end-Triassic catastrophe paved the way for the expansion of the dinosaurs–just as the end-Cretaceous event paved the way for the expansion of mammals.

What Else Do Fossils Tell Us?

As we have seen in this chapter, fossils provide the only direct record we have of past life, so they are vital to our understanding of its history. Moreover, as we found in Chapter 2, fossils are the primary tool used to determine the age of rocks. They also tell us a great deal about past environments (or "paleoenvironments"). For example, fossils of organisms that lived on the sea floor, such as trilobites, give important clues about water depth. If there are only land-derived or terrestrial fossils in a rock, then it was probably not deposited in the sea. This sounds obvious, but it can't be deduced convincingly from any other evidence.

The distribution of fossils can tell us much about past climates. Like modern-day animals and plants, some organisms of the past lived in tropical areas, whereas others lived in temperate or polar regions. Their fossils help us find out what the climate was like in different places; for example,

An early reconstruction of the Iapetus Ocean, based largely on fossil evidence. The trilobite faunas of England, eastern Newfoundland and Nova Scotia (the "Atlantic" faunas) are similar to each other in the Cambrian and Ordovician. Those of Scotland, northwestern Newfoundland and northern New Brunswick and Quebec (the "Pacific" faunas) are also similar to each other, but different from the Atlantic faunas. The best way to explain this distribution of fossils is to propose the existence of a major ocean, the Iapetus Ocean.

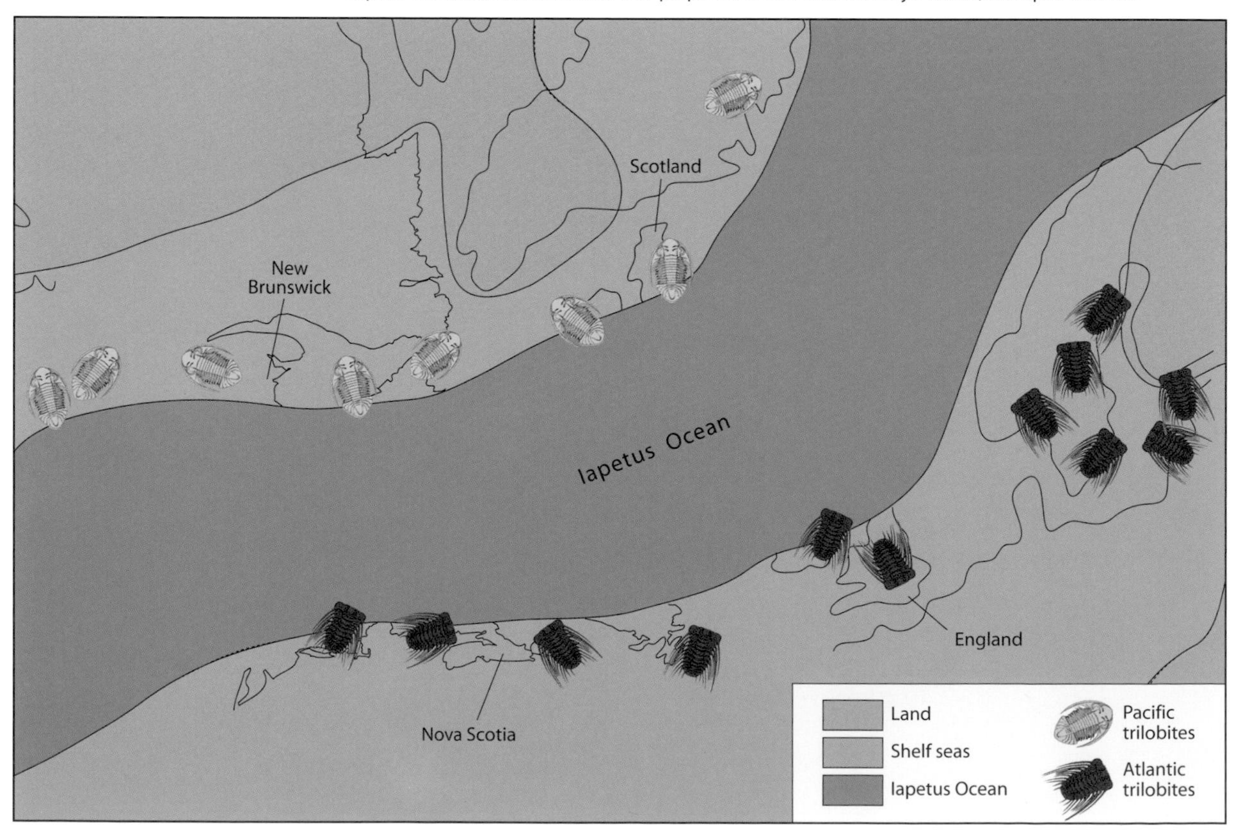

parts of the Canadian Arctic were basking in a warm, temperate climate about 50 million years ago. We know this from the fossils of warm, temperate plants, such as bald cypress trees, found there in rocks of that age.

The distribution of fossils can also help us reconstruct the movement of continents. For instance, the early Paleozoic graptolite and trilobite faunas of the northern British Isles and northern Newfoundland are similar to each other but quite distinct from those of the southern British Isles and southern Newfoundland. This fossil evidence strongly supports the existence of an earlier ocean long before the Atlantic existed.

Fossils can even give us astronomical information. A curious fact learned from early Devonian corals is that, 400 million years ago, a day was about 22 hours long. Daily growth rings found on corals like *Zaphrentis* suggest that a year in the Devonian had about 400 days. Since the Earth's journey around the sun each year has not changed significantly, the length of the day must have increased slowly over geological time because the Earth's rotation has been slowed by tidal friction.

Fossils are not just curiosities but are vital components in the tool boxes of both biologists and geologists. They provide modern life on Earth with a tangible history, and they breathe life into the rocks.

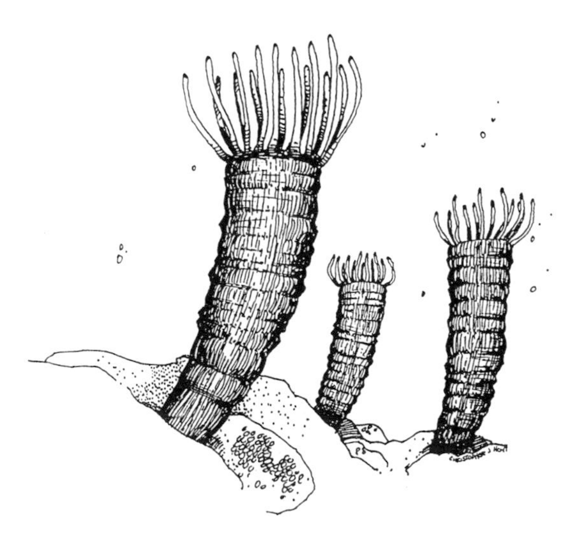

The solitary Silurian-Devonian coral *Zaphrentis*, from Dalhousie, NB and Arisaig, NS.

I've Found a Fossil: Now What?

The thrill of finding a fossil is exciting: you are the first to see the remains of an organism that lived millions of years ago. Your first question is probably "Can I keep it"? All fossils are special, but some are well known to paleontologists, whereas others are rare. You may even be the first person to find the type of fossil that you hold. If you find an exceptional fossil, you should contact your local museum or university geology department. Fossils are a public resource and in some provinces (for example, Nova Scotia) are protected by special laws so they can be saved for study and for the enjoyment of all.

How do you know whether a fossil is exceptional? Fossil bones and skeletons are rare and should always be reported. Fossil footprints, too, are uncommon and important finds. Even the more common fossils, such as fossil plants or shells, might be significant, especially if you find them where fossils have not been found before. Many of the most important fossil finds in recent years have been made by fossil enthusiasts rather than by professional paleontologists. So always be on the lookout–you may just help to put in place another piece in the puzzle of life. And you will be given the credit for the discovery. Remember the van Allen brothers who found the unique *Walchia* forest at Brule!

Some fossil sites are protected—officially or unofficially. Quinn Point, near Jacquet River, NB.

The stromatolite *Archaeozoon* in a late Precambrian sea-shore scene, based on specimens from Saint John, NB. Stromatolites are sedimentary structures produced by the activity of cyanobacteria (organisms that used to be known as blue-green algae). Cyanobacteria trap sediment in algal mats on the sea floor, where the mats tend to build up into mounds, as shown here. Although the Precambrian is not as devoid of fossils as we once thought, stromatolites remain the most obvious evidence of early life.

CHAPTER 4
Into Deepest Time

Supercontinents, Continents and Terranes

In Chapter 1, we introduced the concept of plate tectonics—the division of the Earth's outer layer, or lithosphere, into plates, and the way in which these plates interact along plate boundaries (mid-oceanic ridges, subduction zones and transform faults). Another part of the plate tectonics story is the supercontinent cycle. In this cycle, continents are dispersed, at certain times, over the Earth's surface; at other times, continents come together to form one supercontinent. In the dispersed state, as at the present, several oceans separate the continents. When all the major landmasses are assembled into one supercontinent, there is only one large, surrounding superocean.

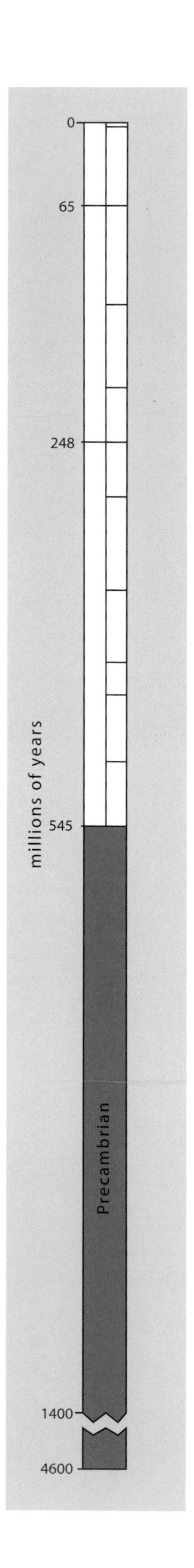

The geological evolution of the Maritimes involves two ancient supercontinents. Rodinia, which existed about a billion years ago, was the earlier of the two. The other was Pangea, which dominated our planet from 350 to 200 million years ago. In this chapter, we show how Rodinia broke apart about 750 million years ago to produce a late Precambrian and early Paleozoic world, with widely scattered continents. In Chapter 5, we describe how those scattered continents came together again to form Pangea, and how this created the Appalachian Mountains and placed the Maritimes at the edge of what was to become the North American continent.

The continental fragments involved in the supercontinent cycle may be thought of as pieces in a giant, ever-changing jigsaw puzzle. When a supercontinent breaks up, continental fragments split apart and oceans form between them. The blocks of land that we know today as continents, such as North America, are larger fragments that were formed as a result of the break-up of Pangea. There are also small displaced blocks of continental crust, or "microcontinents"; Madagascar is a modern example of such a microcontinent. When oceans close, continental fragments reassemble in different patterns. Thus, the size and shape of continents and their distribution over the globe are continually changing.

As an ocean closes through subduction, its heavy oceanic crust sinks beneath the flanking continents. Besides normal oceanic crust, an ocean has volcanic island arcs, seamounts, sedimentary basins and microcontinents. Such features, rather than being subducted like the normal oceanic crust beneath them, may be scraped off and thrust onto the adjacent continental margins. These redistributed bits of continental crust, as well as island arcs and added pieces of oceanic crust, are called "terranes". The successive welding (or "accretion") of terranes to a continent or to each other produces a collage of different kinds of deformed rocks.

The Maritimes represents one such collage of terranes. Most evidence for original relationships of the terranes is destroyed, but the history of the rocks can be deciphered from detailed geological mapping. Geologists examine rock types to determine whether they are igneous, sedimentary or metamorphic. They also search for fossils, which provide evidence for the age of the rocks, as do radiometric dating techniques. In addition to dating the rocks, fossils are used to determine past environments and geography, as we saw in Chapter 3. Mapping also identifies folds and faults, an understanding of which is vital in reconstructing the relative movements of continents and accreted terranes. The chemical composition of samples collected during mapping provides clues about whether rocks originated as parts of oceanic crust, island arcs, or continental fragments. If the last, then samples may provide clues about the larger continental fragment from which they came.

The early geological history of the Maritimes is thus emerging. Many of the details are controversial and will only be resolved with ongoing mapping and with field and laboratory studies of rock samples. But we know enough to present at least part of this dynamic story in this chapter and the next.

An Ancient Piece of Rodinia

The eastern part of the Canadian Shield, in Quebec and Labrador, is composed mainly of rocks that were part of a late Precambrian mountain-building event called the Grenville Orogeny. The Grenville Orogen formed when several continental fragments welded together into the supercontinent Rodinia. The southern boundary of the Grenville Orogen extends across southern Quebec, the Gulf of St. Lawrence, and on to Newfoundland. Unexpectedly, a suite of rocks in what geologists call the Blair River block, near the northern tip of Cape Breton Island, clearly represents a small outlying

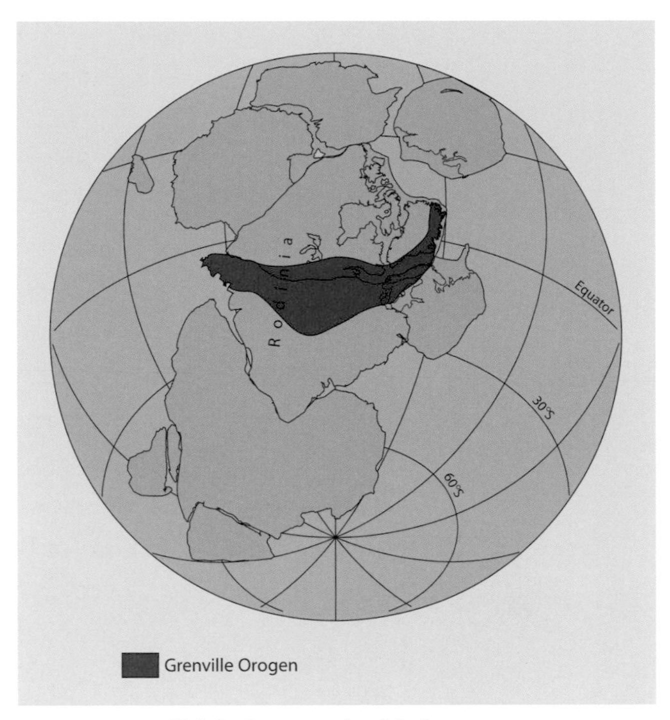

Global paleogeography of the late Precambrian, 850 million years ago.

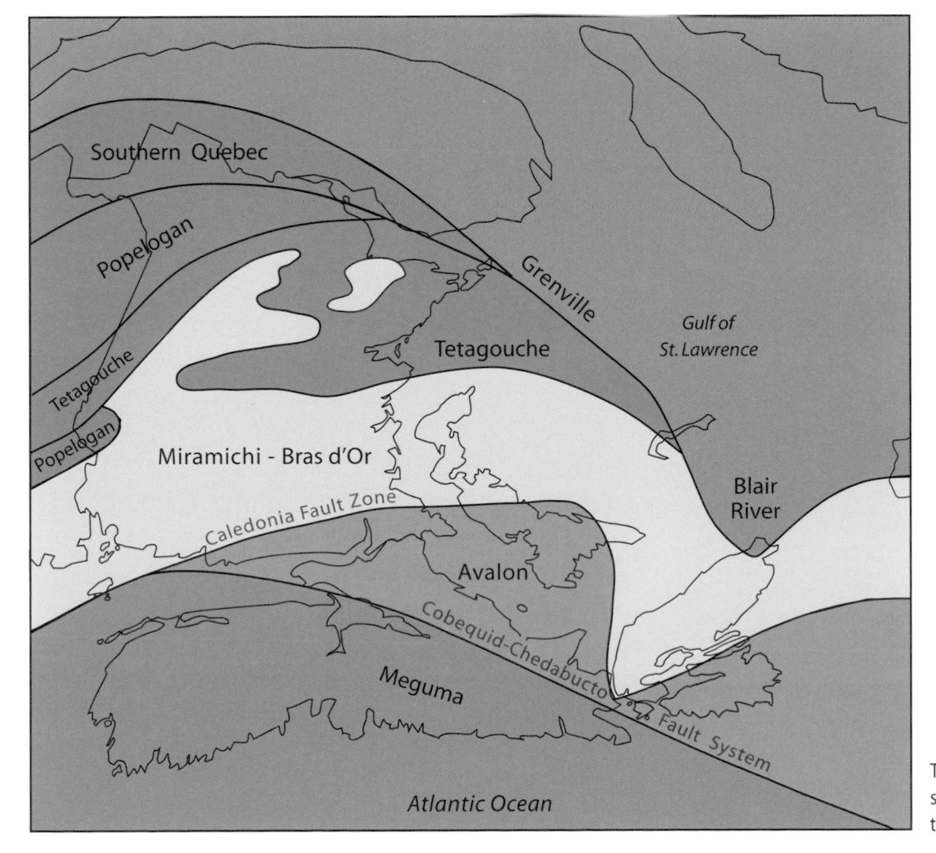

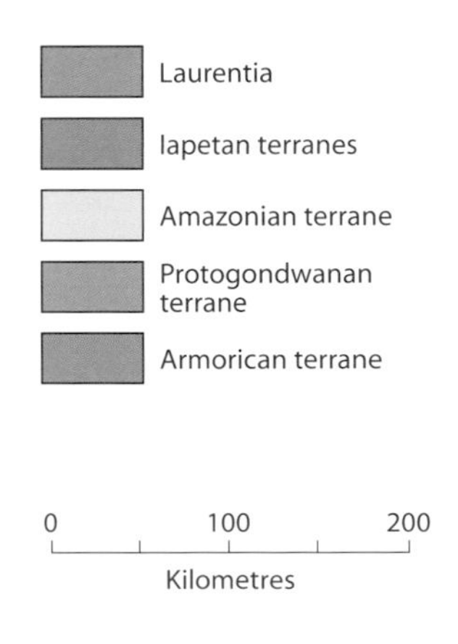

The geological collage of the Maritimes. The terranes, separated by heavy lines, are coloured according to their affinity, as shown in the key.

A Precambrian gneiss from the Blair River area, in northern Cape Breton Island, NS. This gneiss has been dated at 1.5 billion years or more (thus stretching the title of this book).

Metamorphosed banded igneous rocks, just over a billion years old, from the Blair River area in northern Cape Breton Island, NS.

fragment, perhaps a promontory, of the Grenville Orogen and thus also of the Canadian Shield. The rocks of the Blair River block were metamorphosed about a billion (1,000 million) years ago and are the oldest rocks in the Maritimes.

We know that the Blair River block is part of the Grenville Orogen because the rocks and their relationships in these two areas are essentially identical. The rocks in both the Blair River block and the Grenville Orogen include red- and black-banded gneissic rocks. Before metamorphism, these gneissic rocks were mainly igneous rocks that formed as plutons and volcanoes in island arcs between 1,600 and 1,200 million years ago, when Rodinia was beginning to form. Later, in the final stages of the assembly of Rodinia, the gneissic rocks were intruded by distinctive red and white plutonic rocks. Finally, about a billion years ago, all these rocks were metamorphosed once more. Thus the history of Blair River rocks, like that of Grenville rocks, involves two phases of igneous activity and two phases of metamorphism.

The Break-Up of a Supercontinent

During the late Precambrian, the supercontinent Rodinia began to break up into smaller continents, with new oceans developing between them. These new continents and oceans can be thought of as characters in our story, and so deserve an introduction before we continue. One of the larger continental fragments that split from Rodinia was Panamerica, which itself later split into Laurentia, Amazonia, Baltica and Siberia. Another large fragment to split from Rodinia was Protogondwana, encompassing present-day Africa, Arabia, Antarctica, Australia and India. When Amazonia and Protogondwana collided in the Cambrian, about 520 million years ago, they formed the huge southern continent of Gondwana.

The continent of Laurentia encompassed what is now the core of modern North America, represented by the Canadian Shield (including the Grenville Orogen), as well as other regions now separated from North America by the Atlantic Ocean, such as the Hebridean Shield of northern Scotland. Amazonia, as its name suggests, included much of present-day northern South America. Baltica included what is now Scandinavia, the Baltic states and western Russia. Siberia comprised Russia east of the Ural Mountains.

Three ancient oceans played important parts in the early evolution of the Maritimes. The oldest, the Brazilide Ocean, existed between about 750 and 520 million years ago and separated Panamerica from Protogondwana. The Iapetus Ocean formed within Panamerica about 600 million years ago and separated Amazonia from Laurentia; it closed about 430 million years ago. Lastly, the Rheic Ocean opened along the northern margin of Gondwana about 470 million years ago and closed about 350 million years ago, when the assembly of Pangea was almost complete.

Other characters in the story are the terranes. As already mentioned, these were small bits and pieces in and around the three ancient oceans—bits and pieces later welded to the continents. These terranes have been given names, mostly based on places in Atlantic Canada. The largest is the Avalon Terrane, named after the Avalon Peninsula of Newfoundland. This terrane occupies an essentially east-west trending central belt in the Maritimes. The area south of the Avalon Terrane in southern mainland Nova Scotia is occupied by the Meguma Terrane. In the area north of Avalon, between it and Laurentia, there are several other terranes of different ages and origins. Their relationships to one another, and to the Avalon Terrane, are not completely understood. They include the Miramichi-Bras d'Or Terrane of New Brunswick and Cape Breton Island, and the Southern Quebec Arc, Popelogan Arc and Tetagouche Back-Arc terranes of northern New Brunswick.

Most of these terranes originated as volcanic arcs and sedimentary basins along continental margins. Some did not exist until well after the break-up of Rodinia in the late Precambrian. We know this because all rocks exposed in these terranes are younger than Precambrian. The Avalon and Miramichi-Bras d'Or terranes, however, contain Precambrian rocks of continental origin, and this part of the story is further developed in the next section.

Now that we have introduced our cast of characters, we can focus on the break-up of Rodinia. When continents start to break up, rift valleys form—like the modern East African Rift Valley, which suggests that Africa is splitting apart. If the rift valleys continue to widen, sea water washes in and ocean floor spreading begins, as in the Red Sea today. A similar process led to the development of the Brazilide Ocean, as the supercontinent of Rodinia split into the continents of Protogondwana and Panamerica in the late Precambrian. By the

Global paleogeography of the latest Precambrian, 555 million years ago.

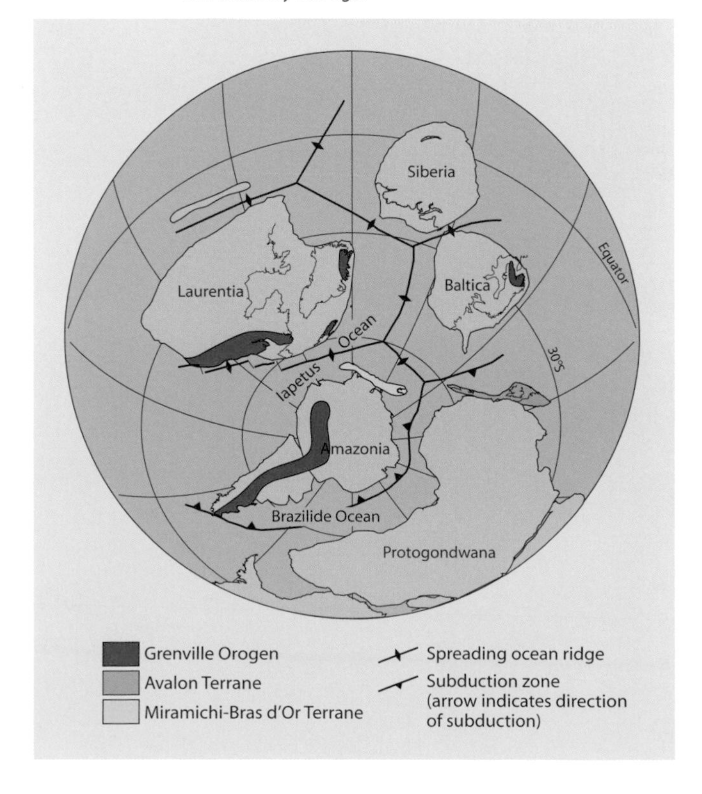

Aerial view of the lighthouse at Louisbourg, NS, which is built on late Precambrian volcanic and sedimentary rocks of the Avalon Terrane.

end of the Precambrian, the Earth had several oceans, and continental fragments formed from different parts of Rodinia had dispersed over the globe. In the Appalachians, the only direct evidence for the initial break-up of Rodinia is a remnant of 760 million-year-old oceanic crust preserved on the Burin Peninsula of Newfoundland.

The Precambrian Core of the Maritimes

The terranes of northern New Brunswick—the Southern Quebec Arc, Popelogan Arc and Tetagouche Back-Arc terranes—are essentially oceanic in origin and have no Precambrian rocks. In southern Nova Scotia, the rocks of the Meguma Terrane probably rest on a Precambrian basement, but no Precambrian rocks are present at the surface. In contrast, the two terranes between northern New Brunswick and southern Nova Scotia, the Avalon and Miramichi-Bras d'Or terranes, have significant Precambrian outcrops. They represent two microcontinental fragments of Precambrian crust and, as such, can be thought of as the ancient core of the Maritimes.

The Avalon Terrane probably originated as a volcanic arc on the margin of the Brazilide Ocean, near the African part of Protogondwana. Today it forms a long, narrow belt of Precambrian rocks, mostly ranging from 740 to 550 million years in age. This belt extends from the Caledonia Highlands

575 million-year-old volcanic rocks of the Avalon Terrane on Guyon Island, off Fourchu, NS. These rocks were deposited during an explosive volcanic eruption. The coloured patches are small volcanic rock fragments.

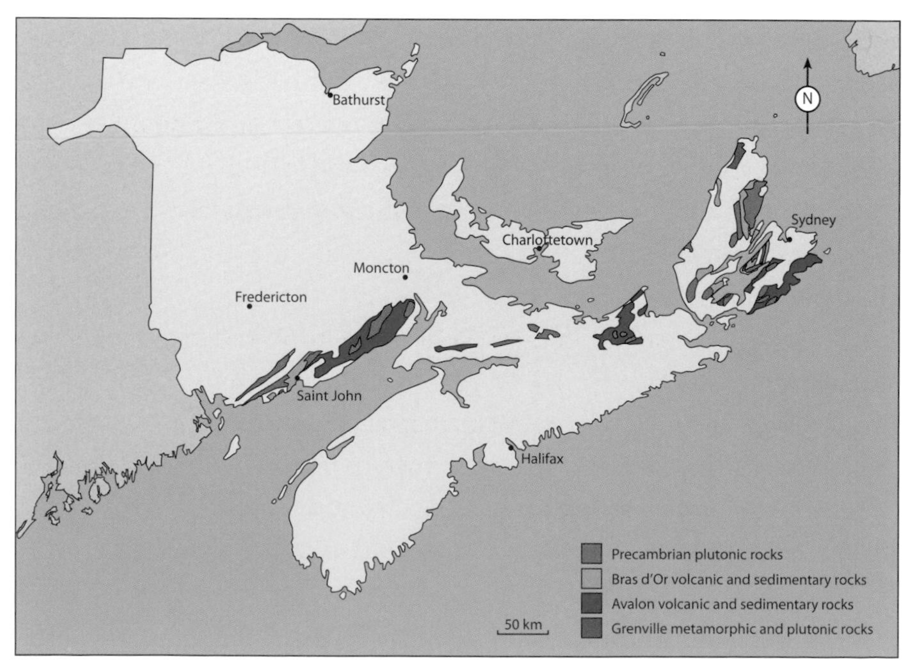

Distribution of Precambrian rocks in the Maritimes.

of southern New Brunswick, through the Cobequid and Antigonish highlands of northern mainland Nova Scotia, to southern Cape Breton Island, and from there to Newfoundland. Similar rocks, also parts of the Avalon Terrane, are found in southern parts of Ireland and Great Britain.

Latest Precambrian siltstone of the Avalon Terrane from Vernons Mountain, east of Saint John, NB. The siltstone, cut by small faults, is made up mostly of volcanic ash deposited in a lake.

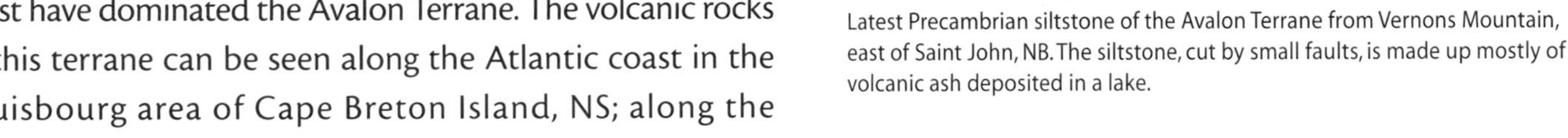

The Avalon Terrane thus had many high, active volcanoes built on thick continental crust, like the modern Andes in South America. In other parts of the world, Mount Pinatubo in the Philippines and Mount St. Helens in the northwestern United States are similar to the huge volcanoes that must have dominated the Avalon Terrane. The volcanic rocks of this terrane can be seen along the Atlantic coast in the Louisbourg area of Cape Breton Island, NS; along the Northumberland Strait north of Antigonish, NS; northwest of Parrsboro, NS, in the Cobequid Highlands; and along the Fundy coast of New Brunswick.

Precambrian rocks of the Miramichi-Bras d'Or Terrane are best exposed north of Saint John, NB, around Rothesay and Green Head Island, and in central Cape Breton Island. In Saint John, the boundary between the Avalon and Miramichi-Bras d'Or terranes is beneath the Reversing Falls Bridge. Precambrian rocks of the Miramichi-Bras d'Or Terrane are also believed to underlie younger rocks in Prince Edward Island and large areas of central New Brunswick and the Gulf of St. Lawrence. Rocks of this terrane probably formed on a passive continental margin, like the margins of the modern Atlantic Ocean. This setting explains the presence of sedimentary rocks deposited in shallow water, including the stromatolite-bearing limestone, now metamorphosed to marble, on Green Head Island in Saint John. Younger Precambrian volcanic arc rocks overlie these older sedimentary rocks, which were metamorphosed and intruded by granite plutons by the early Cambrian. Thus, by that time, the passive margin had become an active, subducting continental margin, probably located at the edge of Amazonia.

By the end of the Precambrian, Rodinia had broken into fragments; its successor, the supercontinent Pangea, was still 200 million years in the future. This scattering of continents meant that there were many continental shelves on which life could experiment. And experiment it did, in one of the most dramatic episodes in the history of the Earth—the Cambrian "explosion" of life.

The Primeval Earth

In the Beginning

To humans, planet Earth seems relatively unchanging. Photographs from space reveal the familiar shapes of continents, blue oceans, and swirling clouds. But this impression of stability is deceptive. Over several billion years, our atmosphere has undergone dramatic changes in composition, and the climate has swung from "greenhouse" to "icehouse" conditions. The mechanism of plate tectonics has moved continents, created mountains and given birth to oceans. Erosion has worn mountain ranges down to their roots. Throughout this upheaval, life has evolved and become increasingly complex. How was the Earth formed, what happened in its early history, and how did life develop?

Earth and the other rocky planets—Mercury, Venus and Mars—began to form more than 4.6 billion years ago. According to the latest ideas, within the developing solar system, cosmic dust and gases lumped together to form larger particles. These particles collided and stuck together (or "accreted") due to gravity and kept on growing until they became planets. Asteroids, meteors and comets represent the leftovers from this process. Impacts at the surface of the early Earth, as well as radioactivity within it, heated our planet and caused it to melt, at least partially. Most of the heavier elements like iron and nickel sank and accumulated at the core, which thus came to be surrounded by lighter elements that eventually formed the mantle and crust.

For hundreds of millions of years after its formation, the Earth continued to be bombarded by meteors and comets. The surface was too hot for anything to survive or for oceans to form, and the atmosphere consisted of a noxious mixture of carbon dioxide, nitrogen and water vapour, with smaller amounts of methane, ammonia and sulphur dioxide. These gases were emitted by volcanoes and represent what was left after the Earth's rocky materials had formed. Oxygen, vital to most modern plants and animals, was absent. Consequently, there was no ozone layer to provide protection against the Sun's deadly ultraviolet rays. By about 3.8 billion years ago, the planet had cooled enough for water vapour to condense and form permanent oceans, a necessary step before life could exist.

The First Life

How did life first develop in the poisonous air and scorching climate of the primeval Earth? The coolest place was in the ocean, and the most nutrient-rich environment was around sea floor vents. However, sunlight could not have penetrated that far, and thus could not have been a source of energy. In the absence of solar energy, certain gases were available, and a process known as chemosynthesis evolved. This is the breakdown of chemical compounds to produce the energy needed for life. Hydrogen sulphide, a gas expelled from volcanoes and deep sea vents, is one of the gases used by organisms to provide energy in these apparently hostile environments.

Perhaps life began around deep sea vents, with chemical reactions producing the necessary compounds, such as amino acids. Alternatively, life may have started when lightning, reacting with the primordial brew of organic-rich compounds in shallow oceanic water, created some of the amino acids. Or perhaps some combination of these events was responsible. Whatever happened, over eons the oceans and seas provided protection from the harmful ultraviolet light of the Sun while letting potentially beneficial sunlight penetrate. Ultimately, organisms harnessed sunlight to produce energy through photosynthesis, a giant leap forward in the evolution of life.

900 million-year-old stromatolites of the genus *Archaeozoon*. Green Head Island, Saint John, NB.

The simplest living organisms known today are prokaryotes (cyanobacteria—formerly known as blue-green algae—and bacteria). A prokaryotic cell is one that does not have a nucleus. The oldest known fossils are prokaryotes from rocks 3.5 billion years old. These were probably cyanobacteria, which can photosynthesize, live in colonies and secrete a sticky substance that sediment sticks to, building up successive layers in mats or mounds called "stromatolites". In most Precambrian rocks, stromatolites are the only fossils visible to the naked eye. Examples of Proterozoic stromatolites are found within the city limits of Saint John. Stromatolites declined about one billion years ago and appear only sporadically in the subsequent geological record. In the Maritimes, examples are found in Carboniferous rocks (about 330 million years old) at Hopewell Cape, NB and in Jurassic rocks (about 200 million years old) near Scots Bay, NS. Today, stromatolites are found only in warm shallow water that is very salty, thus preventing other organisms from grazing on the stromatolite-building cyanobacteria.

Detail from a polished slab of *Archaeozoon* from the Saint John, NB, area. This specimen is displayed in the New Brunswick Museum.

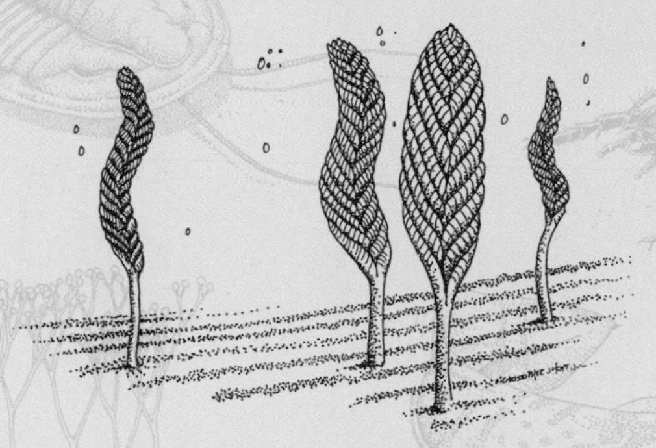

Late Precambrian *Charniodiscus*, an Ediacaran sea-pen-like form.

The Oxygen Revolution and Multicellular Life

By the early Proterozoic, about 2.5 billion years ago, the proliferation of cyanobacteria had "polluted" the Earth's atmosphere with oxygen, a by-product of photosynthesis. Oxygen was toxic to most of the organisms then living. Some scientists now estimate that in a short period of time, geologically speaking (within 100-200 million years), the percentage of free oxygen in the atmosphere rose from essentially zero to about 20 percent of today's level. This major oxygenation is reflected in the rocks. The iron formations of the Canadian Shield represent the rusting (or oxidation) of iron compounds dissolved in the ocean at this time. These formations, worldwide in geographic distribution but restricted largely to the Proterozoic in time, are the primary source of iron for today's steel industry. None are found in the Maritimes.

The oxygen-rich atmosphere was possibly the stimulant necessary for the next major step in the evolution of life: the appearance of eukaryotic cells (cells with a nucleus). Eukaryotic cells are the basis of all multicellular animals and plants, and they led to a new and exciting episode in the Earth's history. The first eukaryotes, like the prokaryotes, were simple, microscopic, single-celled organisms.

The decline in the abundance of stromatolites about one billion years ago roughly coincided with the earliest multicellular animals. This suggests that the evolution of multicellular grazers was responsible for the disappearance of most stromatolites. The rapid development of diverse life would have further increased the level of oxygen and hence the formation of the ozone layer. However, life in the Precambrian remained, primarily at least, in the sea, as the marine environment was more stable and there was less risk from harmful radiation. The earliest evidence of animal macrofossils (that is, those visible to the naked eye) are the "Ediacaran" organisms, found in 600 million-year-old late Proterozoic rocks. Named after the Ediacara Hills in Australia, these organisms are impressions of soft-bodied animals that may have been related to jellyfish, arthropods, molluscs, echinoderms and worms. However, some paleontologists have suggested that Ediacaran organisms were a separate group unto themselves, totally unknown today. No Ediacaran

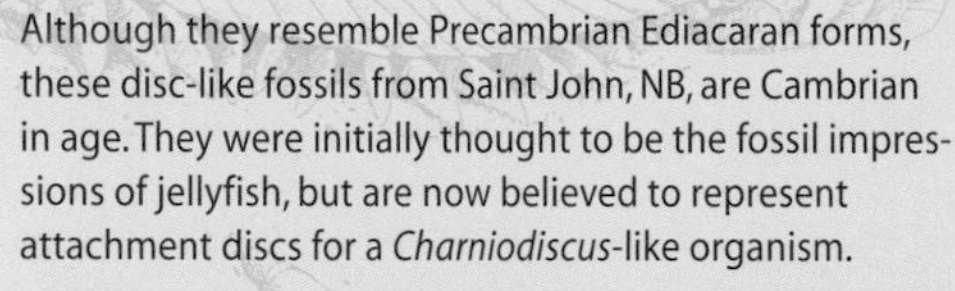

Although they resemble Precambrian Ediacaran forms, these disc-like fossils from Saint John, NB, are Cambrian in age. They were initially thought to be the fossil impressions of jellyfish, but are now believed to represent attachment discs for a *Charniodiscus*-like organism.

fossils are known from the Maritimes, although Ediacaran-like fossils are found in Cambrian sedimentary rocks in Saint John, NB.

Since the Precambrian, rock and fossil records suggest that, although climates have fluctuated, the make-up of the atmosphere has been relatively stable. For most of that time, it is thought that its composition has been about 78 percent nitrogen and 21 percent oxygen, the remaining one percent being a mixture of carbon dioxide, argon, methane, water vapour, ammonia and other trace gases.

"Snowball Earth"

Geologists have long puzzled over the presence of thick suites of Proterozoic rocks derived from glacial deposits, immediately above which are carbonates (limestones and dolomites) that were presumably deposited in a warm ocean. Such suites of strata, about 700 million years old, are found, for example, in the Mackenzie Mountains of northern Canada and in the African country of Namibia. Indeed, they are so widespread that the conditions responsible for these sediments probably extended over the whole Earth.

The "snowball Earth" theory provides a possible explanation for these unusual sequences of rocks. According to this theory, the Earth was covered by ice, due to an extreme cooling event. After the break-up of the supercontinent Rodinia about 750 million years ago, the various new continents all drifted into more or less equatorial positions. Since more energy from the Sun bounces back into space from the land than from the sea, and since equatorial regions receive a greater share of the Sun's energy, the Earth retained less of the Sun's heat while the continents were in this equatorial arrangement. Because of this process, so much heat was reflected back into space that a global deep freeze ensued, and much of the planet became covered with deep glacial ice. If this theory is true, the Maritimes must have been under ice, although there are no known late Precambrian glacial deposits in our region.

Meanwhile, volcanic activity continued unabated, producing large quantities of such greenhouse gases as carbon dioxide. Under normal conditions, these gases become absorbed into the oceans, but during the global glaciations this "sink" was not available, or at least not as large. As a result, gases built up so much in the atmosphere that they triggered a runaway "greenhouse" effect. Not only did it become warm enough for the ice to melt, but tropical conditions ensued globally. Once oceans were ice free, the carbon dioxide that had built up in the atmosphere combined with calcium in sea water to form carbonates, which were precipitated above the glacial deposits.

Scenes like this one help us visualize a possible "Snowball Earth" landscape

As dramatic as this scenario is, there is as yet no better explanation for aspects of the late Precambrian rock record. This "snowball cycle" occurred at least twice, and perhaps three times, within the last few hundred million years leading up to the Cambrian "explosion" discussed in Chapter 5. These dramatic environmental events may have triggered the major evolutionary events around the Precambrian-Cambrian boundary, about 545 million years ago.

During warmer episodes, the Precambrian landscape would have been barren.

A Quick Guide to Some Continents, Oceans and Terranes

The Supercontinents

Rodinia was a supercontinent that included most of the continental areas on Earth from about one billion to 750 million years ago. Rodinia formed as several smaller continents collided to create the supercontinent. These collisions produced mountain ranges that are preserved today as orogens, one of which, the **Grenville Orogen**, now forms the southeastern part of the Canadian Shield. The **Blair River block** in northern Cape Breton is the only piece of the Grenville Orogen exposed in the Maritimes. During the late Precambrian, Rodinia split up into smaller continents, such as Protogondwana and Panamerica. The name Rodinia comes from *rodina*, the Russian word for "homeland".

Pangea was a supercontinent that dominated the Earth about 350 to 200 million years ago, from the time of the Carboniferous coal forests to the dawning of the dinosaurs. It formed from the coming together of Gondwana, Euramerica, and other middle Paleozoic landmasses. Pangea was surrounded by a superocean known as the Panthalassic. The name Pangea comes from Greek and means one Earth or one land. Similarly, Panthalassia means "one sea".

Other Continents

Panamerica formed as Rodinia split up about 750 million years ago and lasted until it broke up into the smaller continents of Amazonia, Laurentia, Baltica and Siberia about 600 million years ago. We call it "Panamerica" because it encompassed much of the ancient cores of both North and South America. Former parts of Panamerica contain remnants of the Grenville Orogen.

Protogondwana, like Panamerica, was formed as Rodinia split up about 750 million years ago. It lasted as a separate continent until the early Cambrian, 520 million years ago, when it collided and amalgamated with Amazonia to form Gondwana. Protogondwana included modern Africa, Arabia, Antarctica, India and Australia.

Amazonia included much of present-day South America and existed as a separate landmass from about 600 to 520 million years ago, when it collided and amalgamated with Protogondwana to form Gondwana.

Laurentia encompassed North America (including Greenland) east of the Rockies, north of the Ozarks and west of the Appalachians. It also included what is today northwestern Scotland, which was separated from the North American part of Laurentia by the opening of the North Atlantic Ocean, between 50 and 100 million years ago. Laurentia existed from about 600 to 430 million years ago, when it collided and amalgamated with Baltica to form Euramerica. Laurentia is named after the Laurentian Highlands in Quebec.

Baltica includes Scandinavia, the Baltic states, and the European part of northern Russia. It was a separate continent from about 600 to 420 million years ago, before it united with Laurentia to form Euramerica.

Euramerica was formed by the collision and amalgamation of Laurentia, Baltica, and several terranes about 420 to 400 million years ago. It is sometimes called the "Old Red Sandstone continent" because its different parts came together in the Devonian, when mainly red sedimentary rocks were being formed. Euramerica lasted until about 350 million years ago, when it joined with Gondwana and other continents to form the supercontinent Pangea.

The name **Gondwana** was originally applied to a huge continent that included today's southern continents (Africa, Antarctica, Australia and South America, as well as the Indian subcontinent) and existed after the break-up of Pangea. However, the term has also come to be used for a large continent of similar (though not identical) composition, a continent that existed prior to the formation of Pangea. It is mostly in this pre-Pangean sense that the name Gondwana is used in this book. This earlier version of Gondwana was formed by the collision and amalgamation of Amazonia and Protogondwana about 520 million years ago. It lasted until about 350 million years ago, when it became part of Pangea. Gondwana means "land of the Gonds", after ancient inhabitants of India.

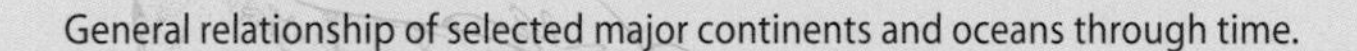

General relationship of selected major continents and oceans through time.

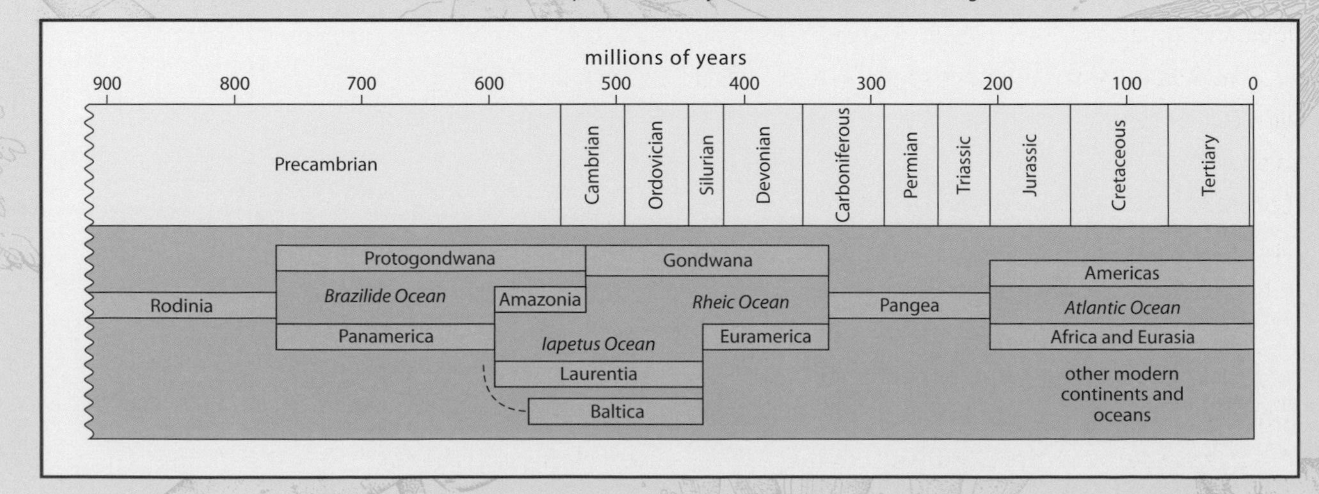

The Oceans

These cliffs at Cape St. Mary, NS, are formed of metasedimentary rocks of the Meguma Terrane

The **Brazilide Ocean** opened as Rodinia broke up about 750 million years ago. It had Panamerica on its northern margin and Protogondwana on its southern margin. The Brazilide Ocean closed when Amazonia and Protogondwana came together to form Gondwana about 520 million years ago.

The **Iapetus Ocean** opened as Panamerica split into Laurentia, Amazonia, and Baltica about 600 million years ago. In the Maritimes, this ocean existed until about 430 million years ago, when several terranes collided and joined with Laurentia and Baltica to form part of Euramerica. In Greek mythology, the god Iapetus was father of Atlas, after whom the Atlantic Ocean is named.

The **Rheic Ocean** opened between the Avalon Terrane and Gondwana about 470 million years ago. Within the Rheic Ocean was the Meguma Terrane, which split from Gondwana early in this ocean's development. Part of the Rheic Ocean closed about 390 million years ago, when the Meguma Terrane collided and amalgamated with the Avalon Terrane to become part of Euramerica. The remaining part of the Rheic Ocean closed about 350 million years ago when Gondwana united with Euramerica to form Pangea. The Greek goddess Rhea was the daughter of Oceanus.

The Terranes

The **Meguma Terrane** of southern mainland Nova Scotia consists largely of Cambrian to Ordovician sedimentary rocks deposited along the passive continental margin of Gondwana. It is referred to as an **Armorican terrane** because it was probably part of the Armorican microcontinent, which separated from Gondwana about 470 million years ago. The Meguma Terrane eventually collided and amalgamated with Euramerica about 390 million years ago.

The **Avalon Terrane** consists mostly of late Precambrian rocks associated with volcanic arc settings on the Protogondwanan continental margin and, for this reason, is referred to as a **Protogondwanan terrane.** It split away from Gondwana about 480 million years ago, colliding and amalgamating with Laurentia about 400 million years ago. It is named after the Avalon Peninsula in Newfoundland, where it is well exposed.

The **Miramichi-Bras d'Or Terrane** includes older Precambrian sedimentary rocks (now partly metamorphosed) that were deposited on a passive continental shelf. Later Precambrian and early Cambrian volcanic arc rocks indicate subsequent development of an active continental margin. The Miramichi-Bras d'Or Terrane may have originated off the continental margin of Amazonia and is thus referred to as an **Amazonian terrane**. It collided and fused with Laurentia about 430 million years ago, during the closure of the Tetagouche Back-Arc Basin.

The **Popelogan Arc Terrane** formed from an Ordovician volcanic island arc in the Iapetus Ocean off the northwestern margin of the Miramichi-Bras d'Or Terrane in New Brunswick.

The **Tetagouche Back-Arc Terrane** formed in northern and western New Brunswick when the Popelogan Arc split in two in the later part of the Ordovician.

The **Southern Quebec Arc Terrane** formed as a Cambrian-to-Ordovician volcanic island arc in the Iapetus Ocean, off the continental margin of Laurentia.

The **Kingston Arc Terrane** formed as a Silurian volcanic arc on the margin of the Miramichi-Bras d'Or Terrane. This was a late developing terrane and so is not shown on the terrane map.

The relationship of continents, oceans and terranes (including microcontinents, volcanic arcs and back arcs) that produced the Maritimes collage during the Late Precambrian and Paleozoic.

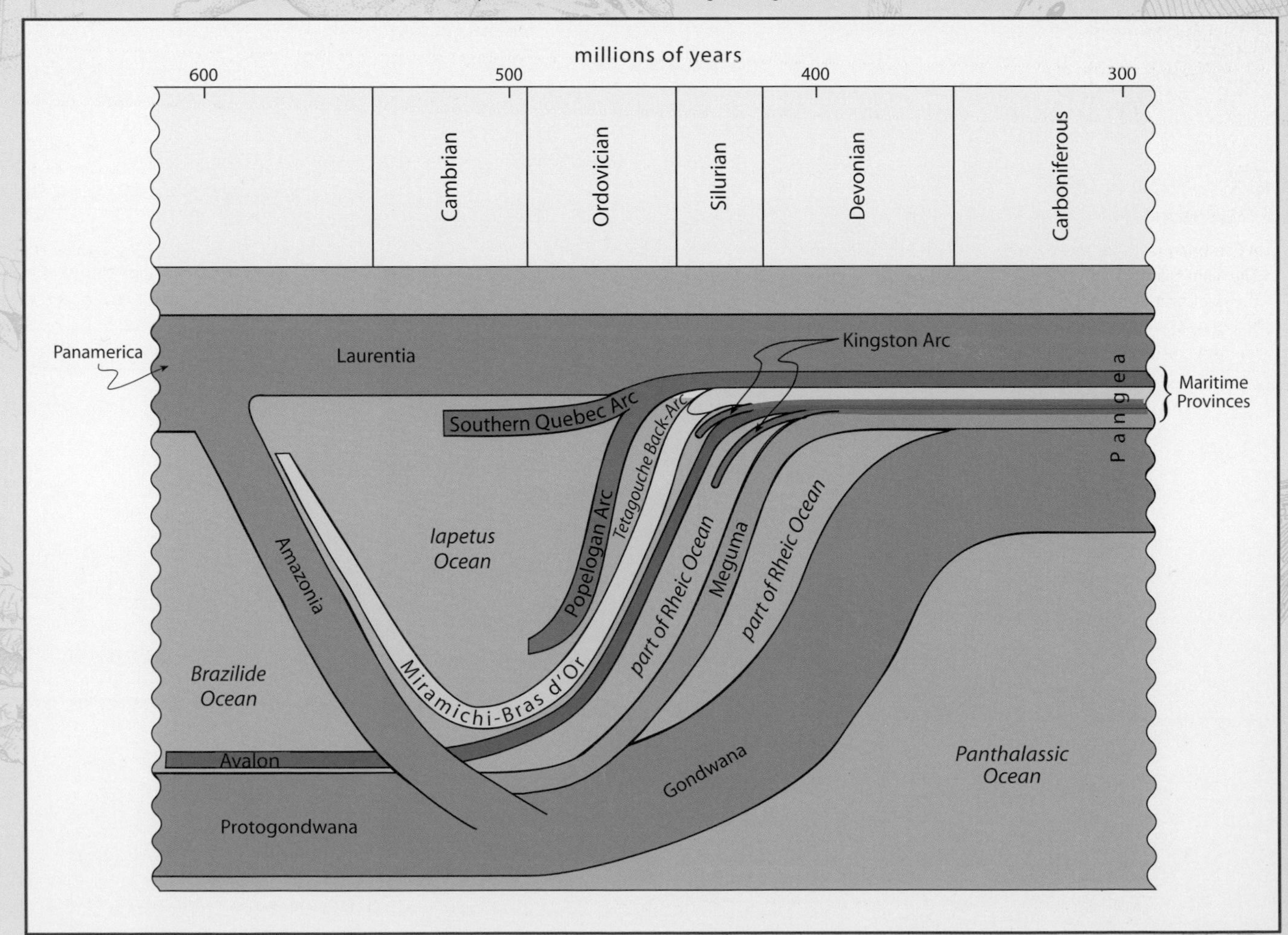

CHAPTER 5
The Pieces Come Together

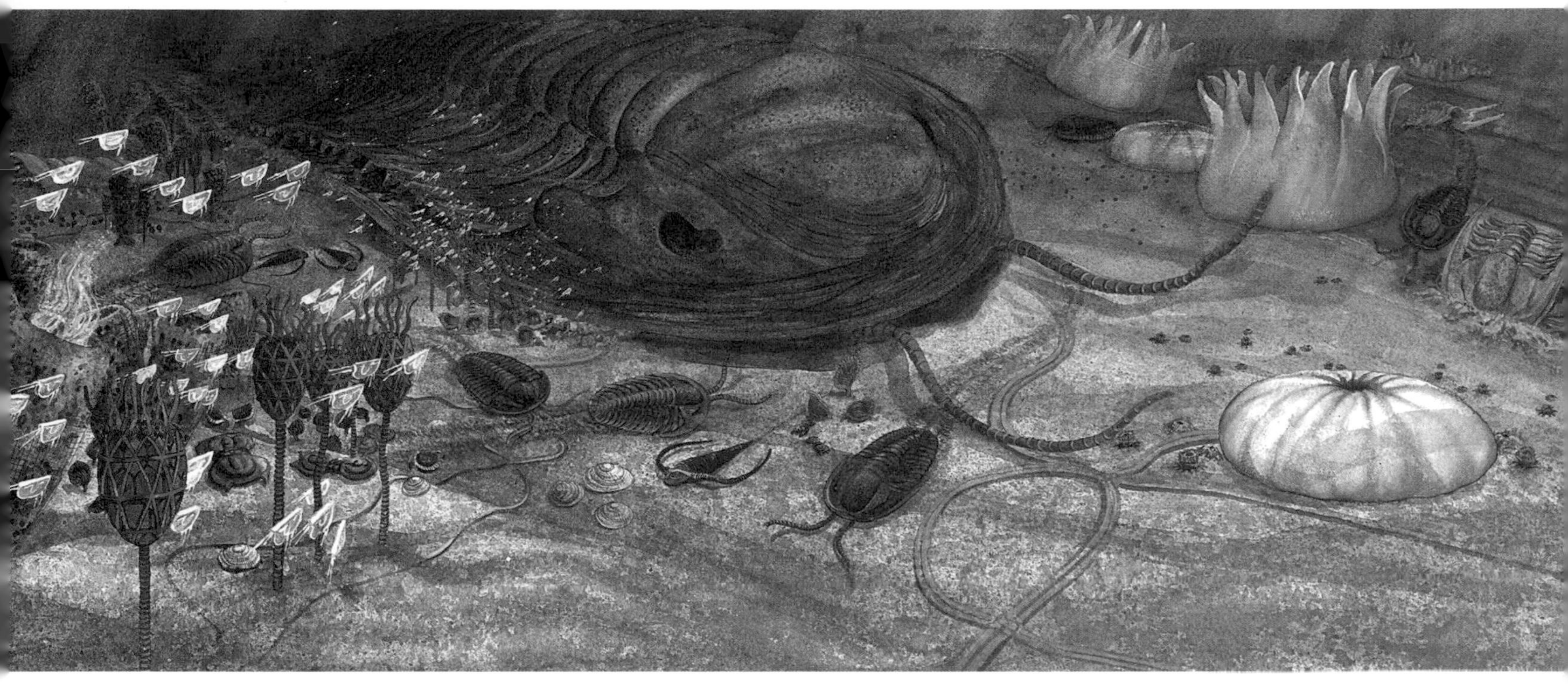

A Cambrian sea-floor scene based on fossils found around Saint John, NB. The main characters are the large trilobite *Accadoparadoxides*, some "jellyfish" to the right, and, to the left, several red eocrinoids of the genus *Eocystites*. Other trilobites are the black *Conocoryphe*; the smaller brown *Solenopleura*; and, at the far right, the larger, brown, rolled-up (molting) *Ctenocephalus*. Hiding among the eocrinoids are small *Agnostus*-like trilobites. The reddish, spiny trilobite-like creature in front of *Accadoparadoxides* is *Hyolithes*. Small, translucent, white ostracods float past the eocrinoids, behind which are brown sponges of the genus *Protospongia*. About to be squashed by the advancing *Accadoparadoxides* are specimens of the brachiopod *Orthis*, a form with hinged shells. The trackway in the foreground is *Psammichnites*, the trace maker being suitably out of the scene as we don't know what it looked like.

The Cambrian "Explosion"

One of the most dramatic events in the history of life was the Cambrian "explosion", the great diversification of animal life, about 545 million years ago, around the Precambrian-Cambrian boundary. Once believed to have been instantaneous in a geological sense, this event is now known to have taken place over about ten million years.

Life in the latest Precambrian seas was becoming increasingly varied, but the animals were mostly soft-bodied, so were rarely preserved. In the Cambrian "explosion", a bewildering array of animals with hard skeletons became common for the first time. The earliest harbinger of change, at the very end of the Precambrian, was an increase in the diversity of trails and shallow burrows, reflecting an increase in the number and

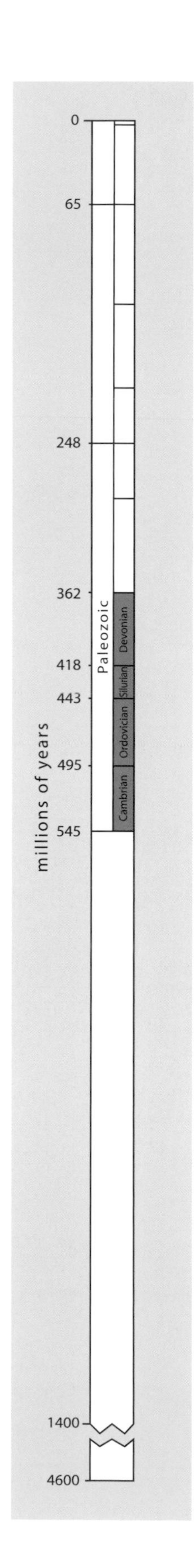

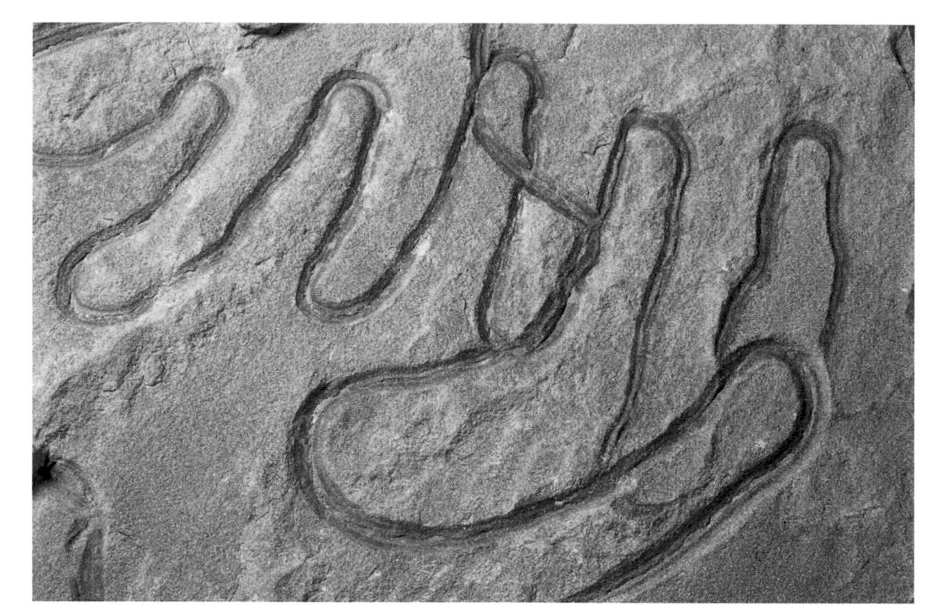

The trace fossil *Taphrhelminthoida* from early Cambrian strata of the Avalon Terrane, near St. Martins, NB. This trace was produced as an animal burrowed through the sediment searching for food.

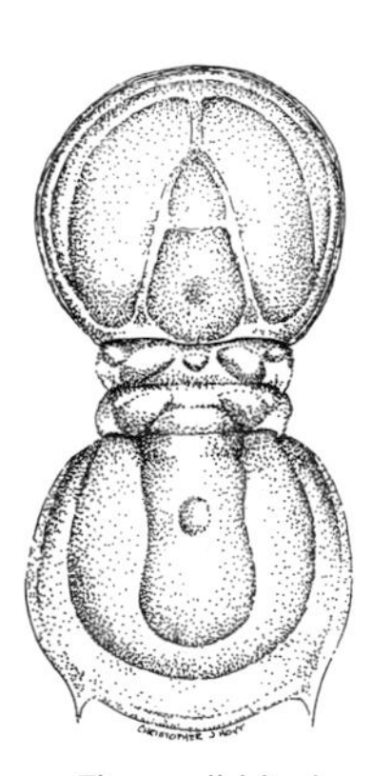

The small, blind Cambrian trilobite *Agnostus*, from Saint John, NB.

activity of small animals living on the sea floor. These trace fossils were more complex than their precursors, reflecting more complicated behaviour. This was followed by an increase in the variety and abundance of "small shelly fossils". Deep, vertical burrows and larger shelled fossils came next, including the oldest brachiopods and bivalves. Snails appeared and left winding paths across the Cambrian sea floor, and archaeocyathids (coral-like organisms) built reefs. There were also many strange forms that later became extinct, such as worm-like creatures covered with shelly plates. The final stage of the Cambrian "explosion" was marked by the appearance of trilobites, which left distinctive trails, burrows and shells in the fossil record. Trilobites were among the earliest animals known to have eyes. Thus, for the first time, some forms of life could see the world in which they lived.

In the Maritimes, the first stage of the Cambrian "explosion" is preserved near Saint John, NB, where there are trace fossils and so-called "small shelly fossils", such as *Hyolithes*, *Sabellidites* and the early snail *Aldanella*. Similar deposits and fossils are found north of Antigonish, NS, and in southeastern Cape Breton Island.

The early Cambrian small shelly fossil, *Hyolithes*.

Oceans Come and Oceans Go

As Cambrian life was burgeoning, the process of plate tectonics continued to modify the shapes of continents and oceans. At the beginning of the Cambrian, the pieces of what would become the Maritimes were still dispersed or yet to be created. Over the next 150 million years, these pieces gradually came together, although in a somewhat erratic fashion. Indeed, some pieces scattered even farther afield before finally assembling.

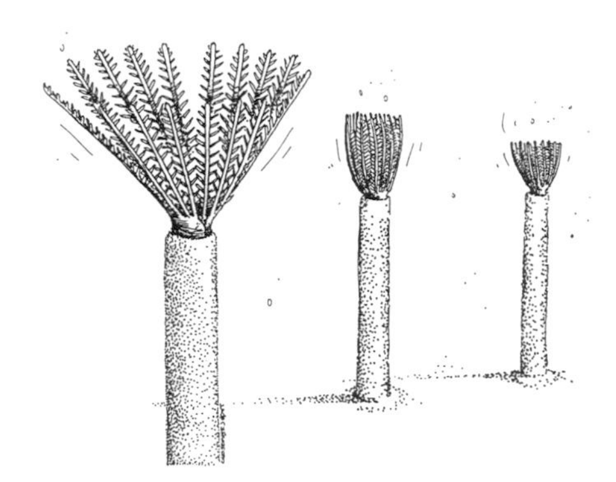

The early Cambrian tube-like *Sabellidites*.

A new ocean, the Iapetus Ocean, began to open about 600 million years ago and continued to widen as the old Brazilide Ocean narrowed. The development of the Iapetus Ocean was largely responsible for the break-up of Panamerica, giving rise to several smaller continents, including Laurentia and Amazonia. It was the collision of Amazonia with Protogondwana in the early Cambrian that signalled the final closure of the Brazilide Ocean and the formation of the large southern continent of Gondwana.

The early Cambrian snail *Aldanella*.

Extensive passive margins had developed on either side of the Iapetus Ocean by Cambrian time. Suites of sedimentary rocks from both sides of Iapetus are preserved in Atlantic Canada, and it is clear from rock types and fossils found in each suite that these were deposited, literally, an ocean apart. Cambrian and Ordovician limestone beds that formed in warm, shallow coastal waters on the northern (Laurentian) margin of Iapetus are preserved in western Newfoundland and along the St. Lawrence River near Quebec City. These contrast dramatically with sandstones and shales deposited in colder waters on the southern (Gondwanan) margin of Iapetus. Likewise, fossils found in sedimentary rocks on the Gondwanan side of Iapetus are unlike those of similar age on the opposing Laurentian margin.

Global paleogeography of the Cambrian, 515 million years ago.

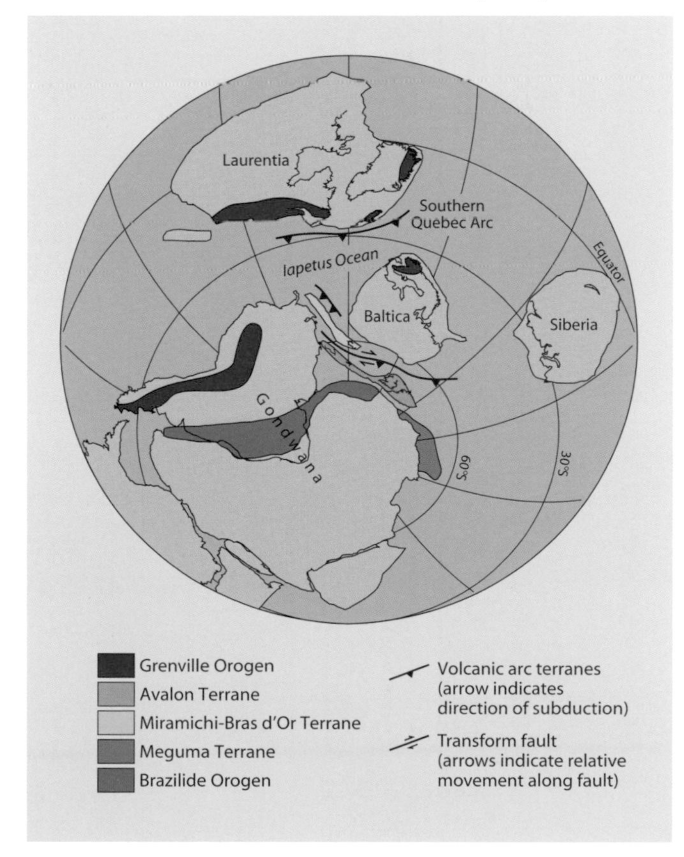

The sediments deposited on the Gondwanan margin of the Iapetus Ocean were like those of the present-day Atlantic margin of eastern Canada—mainly sands and muds eroded from the continents and deposited on the continental shelf. Turbidity currents then carried most of this sediment from the continental shelf to the deep ocean floor. Sedimentary rocks that were likely formed in this way are exposed in the Miramichi Highlands of New Brunswick, in a belt extending northeast from Woodstock to Bathurst. These quartz-rich sandstones and shales of the Miramichi-Bras d'Or Terrane were deposited during the Cambrian and early Ordovician, between 545 and 485 million years ago. The many granitic

Late Cambrian sedimentary rocks of the Miramichi-Bras d'Or Terrane (supporting ATVs), Nepisiguit River near Bathurst Mines, NB.

batholiths that later intruded into these sedimentary rocks show that this terrane is underlain by continental crust (as granite does not form in oceanic crust). Other geological evidence, such as isotope chemistry and radiometric age-dating, suggest that the Miramichi-Bras d'Or Terrane originally lay next to Amazonia. It is thus referred to as an Amazonian terrane.

The Avalon Terrane in the Cambrian was situated on the African margin of Gondwana, which, prior to the closing of the Brazilide Ocean, was part of Protogondwana. The boundary between the Protogondwanan Avalon Terrane and Amazonian Miramichi-Bras d'Or Terrane to the north is presently the site of a major fault system, the Caledonia Fault Zone, along which it is thought that these two terranes, once widely separated, slid together. This movement took place over tens of millions of years, continuing until the Carboniferous Period, about 350 million years ago.

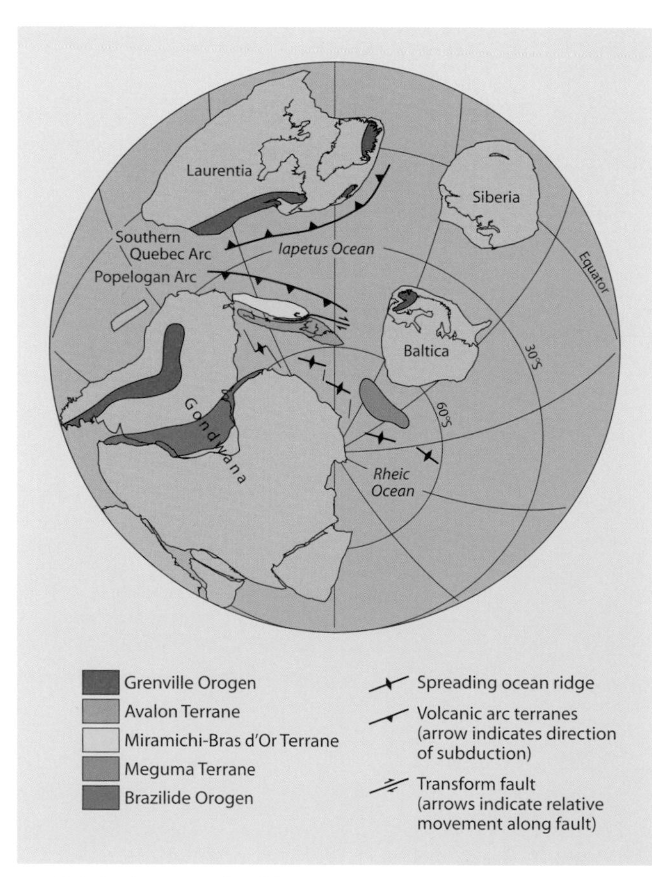

Global paleogeography of the early Ordovician, 475 million years ago.

Volcanic Arcs and the Closing of the Iapetus Ocean

In the middle to late Cambrian, the Iapetus Ocean became more like the modern western Pacific than the present-day Atlantic, with crust being subducted at the ocean's margins. Chains of volcanic islands (volcanic island arcs, like the Philippines and Japan today) formed above the subduction zones. One of the volcanic arcs that developed near the Laurentian margin of Iapetus was the Southern Quebec Arc. Volcanic arcs tend to be accreted to continental margins as underlying oceanic crust is destroyed by subduction. Fragments of the Southern Quebec Arc Terrane are preserved because they were thrust over the Laurentian continental margin during closure of part of the Iapetus Ocean in the Ordovician.

Pieces of another Ordovician volcanic arc, the Popelogan Arc, formed on the Gondwanan side of the Iapetus Ocean. These pieces are preserved as the Popelogan Arc Terrane, which underlies a westerly-broadening belt from the shore of Chaleur Bay into Maine. Most of this terrane is hidden beneath a cover of younger sedimentary rocks, but it is exposed near the Popelogan River, 20 kilometres south of Campbellton, NB, and on Eel River, southwest of Woodstock, NB.

There are some terms and processes related to volcanic arcs that we must briefly introduce in order to explain the next part of our story. By definition, the subduction zone is on the fore-arc side of a volcanic arc. The area that is on the opposite side of (or "behind") a volcanic arc is called a back-arc region. Rift basins commonly open up in back-arc regions, and are thus called back-arc basins. The Sea of Japan, which opened behind the volcanic arc that is Japan, is a modern example of such a back-arc basin.

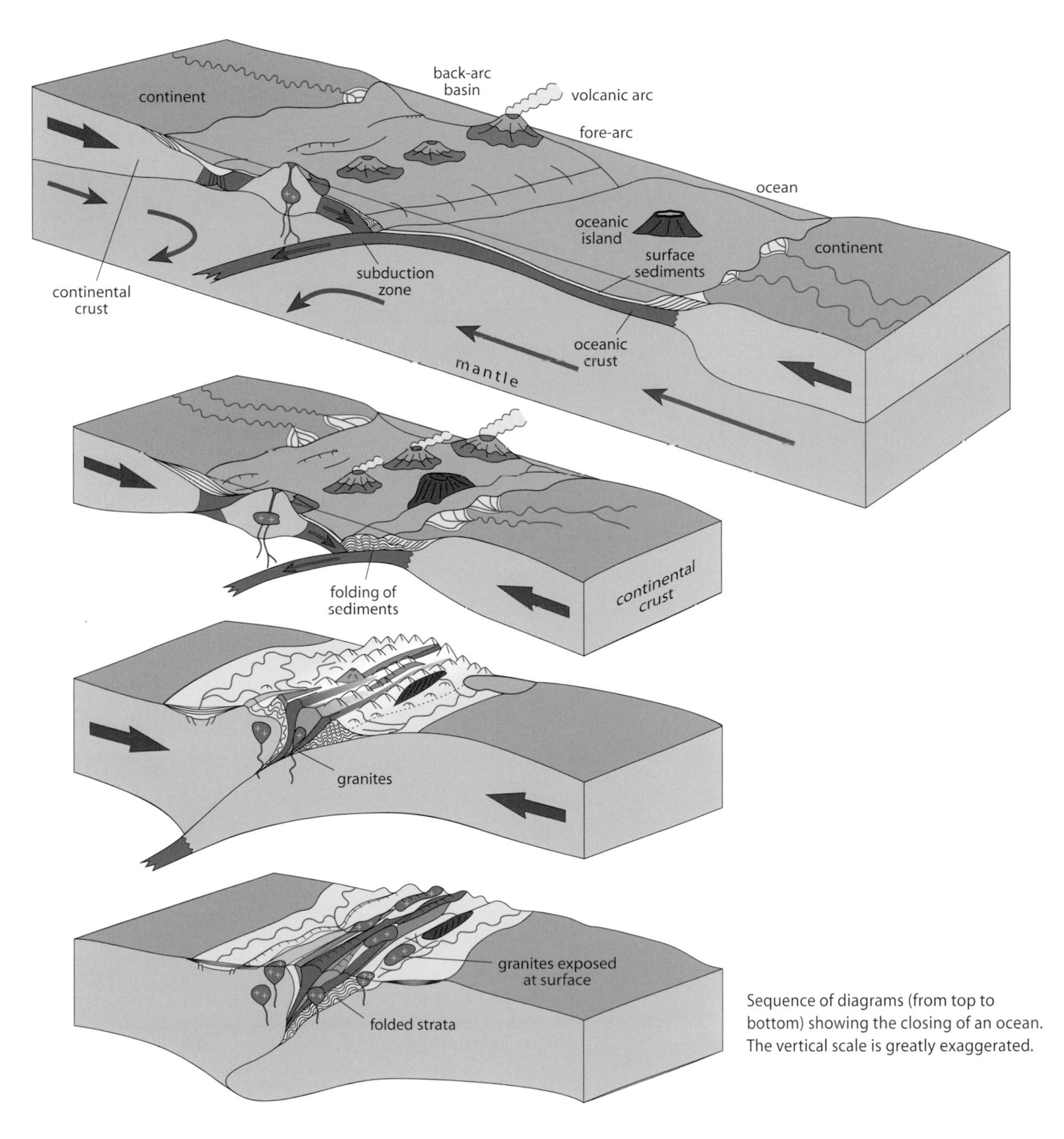

Sequence of diagrams (from top to bottom) showing the closing of an ocean. The vertical scale is greatly exaggerated.

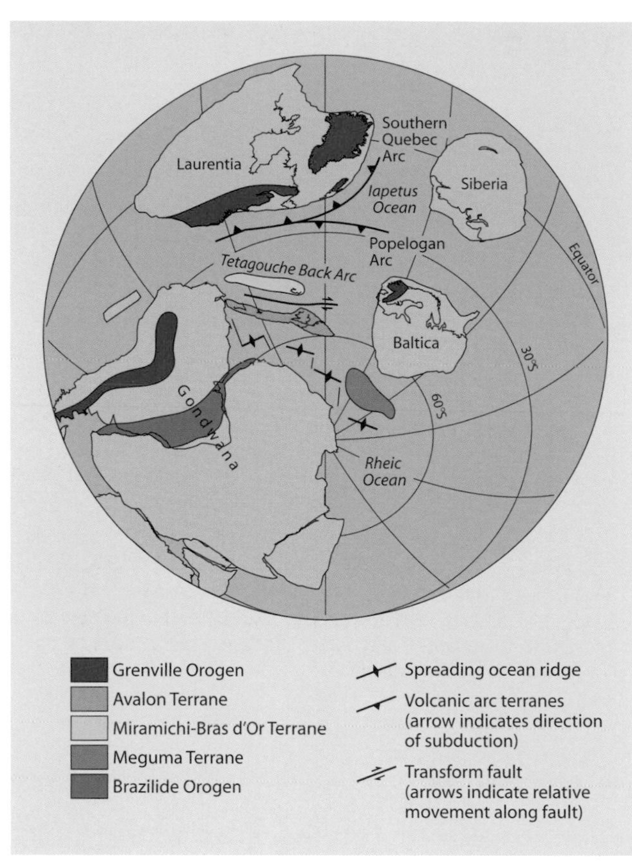

Global paleogeography of the middle Ordovician, 465 million years ago.

During the early to middle Ordovician, a rift basin developed behind the Popelogan Arc in continental crust of the Miramichi-Bras d'Or Terrane. This was the Tetagouche Back-Arc Basin, now preserved as the Tetagouche Back-Arc Terrane. Because the Tetagouche Back-Arc Basin formed on continental crust, large volumes of felsic magma were generated. The resulting volcanic eruptions produced great thicknesses of rhyolitic volcanic lava and ash that were deposited on the sedimentary rocks of the Miramichi-Bras d'Or Terrane. These eruptions also produced most of the lead-zinc deposits in the Bathurst area of New Brunswick.

Rhyolitic volcanic rocks were succeeded by alkali-rich basaltic lavas, including pillow basalts such as those exposed near Bathurst. Alkali-rich basalts are generally produced by "hot spots" within the Earth's mantle and are usually independent of plate boundaries and oceanic crust production, where alkali-poor basalts are the norm. Hot spots today are responsible for the Hawaiian Islands within the oceanic Pacific Plate, and for the Rio Grande Rift in New Mexico within the North American Plate. Evidently, during the Ordovician, there was a hot spot beneath the Miramichi-Bras d'Or Terrane.

By about 460 million years ago, in the middle Ordovician, new oceanic crust had begun to form within the northern part of the Tetagouche Back-Arc Basin as the Popelogan Arc split in two. The basaltic lava that erupted onto the newly forming ocean floor at this time is alkali-poor,

The gorge of the Nepisiguit River at Bathurst Mines, NB, cuts through metamorphosed Ordovician volcanic rocks of the Tetagouche Back-Arc Terrane.

Pillow basalts of late Ordovician age, Tetagouche Back-Arc Terrane near Bathurst, NB. Such pillow basalts are produced by lavas flowing on the sea floor, in this case the floor of the Tetagouche Back-Arc Basin. The sea water cools the outermost crust of lava, but the lava within this crust is still molten and breaks through to form another tongue. This process, repeated again and again, produces the pillow structures.

and is therefore similar in composition to that found on present-day mid-oceanic ridges. Remnants of oceanic crust from the Tetagouche Back-Arc Terrane are exposed near Pointe Verte, NB.

By the late Ordovician, much of the Iapetus Ocean was destroyed, and terranes within the ocean were colliding with Laurentia. All that remained of Iapetus in our region was the Tetagouche Back-Arc Basin, which may have been about 1,000 kilometres wide. This basin was also gradually subducted and was completely closed by about 430 million years ago, in the early Silurian. As the back-arc basin closed, rocks that had been sinking into the subduction zone were forced back to the surface and thrust onto the Miramichi-Bras d'Or Terrane. Now exposed to the west and southwest of Bathurst, NB, these rocks include "blueschists", which are formed from the metamorphism of iron and magnesium-rich volcanic rocks, under very high pressures in subduction zones. Blueschist gets its name from its characteristic bluish tinge, reflecting its mineral composition.

These rocks, near the mouth of the Elmtree River, near Petit Rocher, NB, are preserved Ordovician oceanic crust from the Tetagouche Back-Arc Basin. They are a series of vertical, or "sheeted", dykes that pushed up into the crust during sea-floor spreading. Such rocks are usually destroyed during subduction.

In northern Cape Breton Island, the Miramichi-Bras d'Or Terrane lies directly against the Blair River block of Laurentia. There are no rocks from the once-intervening Ordovician oceanic arc terranes as seen in northern New Brunswick. This may be because, in what is now Cape Breton, the collision was between two peninsulas or promontories—one on the margin of Laurentia and the other on the margin of the Miramichi-Bras d'Or Terrane. The intense compression caused by the collision of these opposing promontories created the high temperature and pressure needed to produce the metamorphic rocks that make up the Miramichi-Bras d'Or Terrane in Cape Breton. Any oceanic arc terranes between the Miramichi-Bras d'Or Terrane and Blair River block on Laurentia would have been thrust up and subsequently completely eroded.

Thus, what was once the mighty Iapetus Ocean is now represented in the Maritimes by a highly deformed belt of rocks that includes the Southern Quebec Arc, the Popelogan Arc and the Tetagouche Back-Arc terranes. These terranes became wedged between Laurentia and the Miramichi-Bras d'Or Terrane as Iapetus closed and are together called the Northern

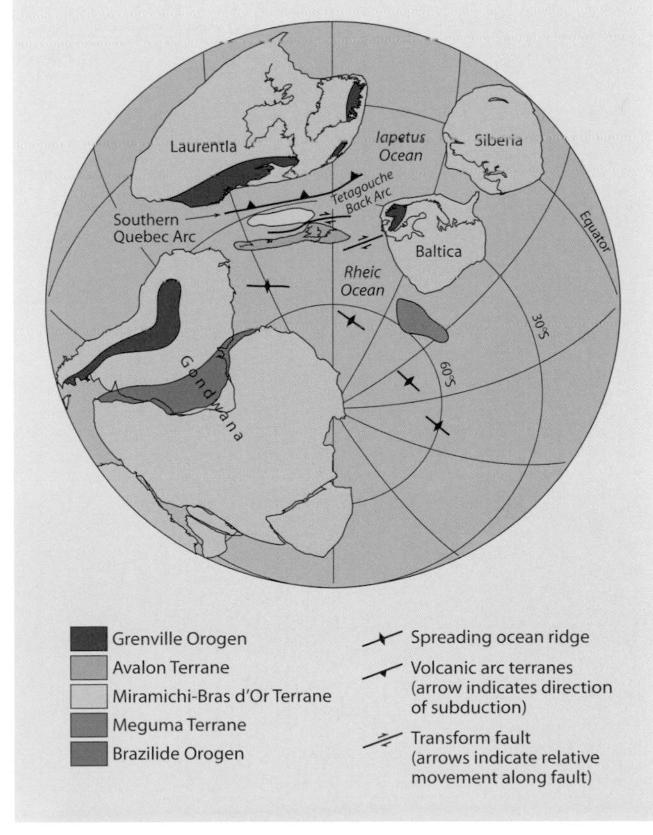

Global paleogeography of the late Ordovician, 455 million years ago.

Gneiss of early Paleozoic age on the bank of the Middle Aspy River, near Cape North, NS, part of the Miramichi-Bras d'Or Terrane.

Acadian Orogen. As we saw earlier, the orogeny that produced the Northern Acadian Orogen was a complex series of collision events, all of which contributed to the building of the ancient Appalachian Mountains over many millions of years.

Northern Interlude

With the closure of the Iapetus Ocean in the latest Ordovician, the focus of plate tectonic activity in the Maritimes shifted south, to the other side of the Avalon Terrane. But late Ordovician, Silurian and Devonian times in the area of the former Iapetus Ocean were by no means devoid of events, as the geological record shows. Before heading south, therefore, the later story of the Northern Acadian Orogen must be told.

During the late Ordovician, well-bedded limestones, commonly referred to as "ribbon rock", were deposited in deep sedimentary basins that remained after the Popelogan Arc had accreted onto Laurentia. Such rocks are best seen in the gorge formed by the Saint John River at Grand Falls, NB.

The Silurian to early Devonian interval is represented by volcanic rocks and fossiliferous sedimentary rocks in the Chaleur Bay and Tobique River areas of northern and central New Brunswick, on the northwestern margin of the Miramichi-Bras d'Or Terrane. These rocks were apparently deposited in faulted-bounded basins developed during final closure of the Iapetus

The gorge of the Saint John River at Grand Falls, NB, cuts through late Ordovician to early Silurian sedimentary rocks (shaly limestone) covering the Popelogan Arc Terrane.

Folded Silurian sedimentary rocks near Jacquet River, NB, deposited after fusion of the various terranes in northern New Brunswick; this fusion produced the Northern Acadian Orogen.

Ocean. Volcanic activity in these basins has left the Maritimes with some of its most striking topographic features. For example, New Brunswick's Mount Carleton (at 820 metres, the highest peak in the Maritimes) consists of hard, silica-rich volcanic rocks that were erupted during the early Devonian. Sugarloaf Mountain in Campbellton, NB, and Bald Peak near Plaster Rock, NB, are also volcanic remnants from this time.

Fossils in early Devonian limestones on the Laurentian margin in the Gaspé Peninsula of Quebec and in the Miramichi-Bras d'Or Terrane in New Brunswick are similar, confirming that these two areas were relatively near each other, no longer an ocean apart.

Another result of the collision brought about by the closing of the Iapetus Ocean was the generation, in the late Silurian and early Devonian (about 420-400 million years ago), of large volumes of magma deep in the thickened crust and underlying mantle. These magmas mostly solidified into plutons beneath the surface, forming bodies of coarsely crystalline rocks, including silica-poor gabbro and silica-rich granite. Uplift and erosion have since exposed many of these plutons at the surface, such as the Pokiok Batholith to the southeast of Woodstock, NB.

By the middle Devonian (about 390 million years ago), most of northern and western New Brunswick was above sea level, and sediment deposition was restricted to river valleys. The plant and fish-bearing beds, exposed along Chaleur Bay between Campbellton and Dalhousie, were probably deposited at the mouth of a tidal river. The fossils from these beds will be discussed later in the chapter.

Sugarloaf Mountain, just outside Campbellton, NB, is the eroded remnant of a Devonian volcano.

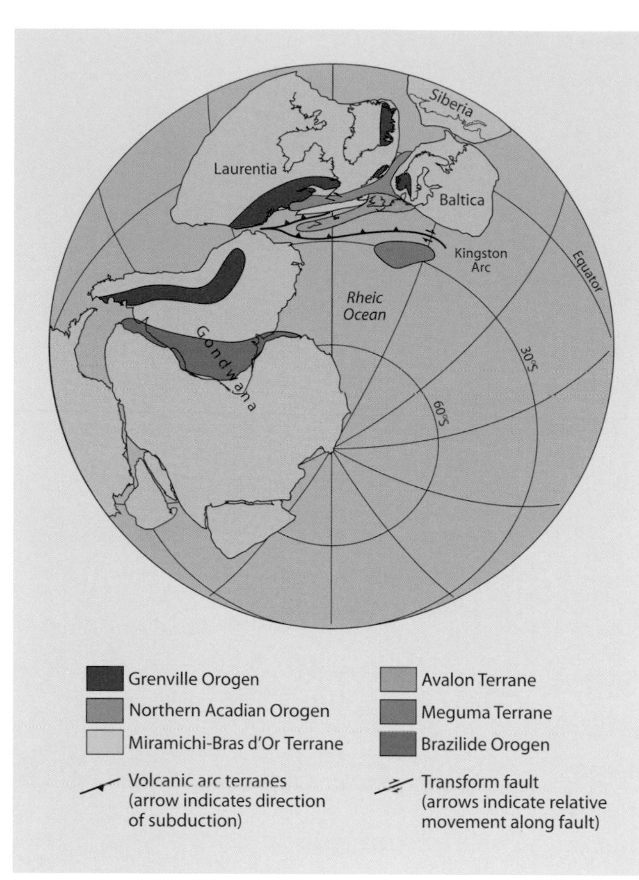

Global paleogeography of the early Silurian, 435 million years ago.

Another Open-and-Shut Case

Just as the closure of the Brazilide Ocean was followed by the opening of the Iapetus Ocean, so, as Iapetus gradually closed toward the end of the Ordovician, the Rheic Ocean opened between Laurentia and Gondwana. Laurentia had by then expanded to include the accreted terranes of the North Acadian Orogeny. As the Rheic Ocean widened throughout the Ordovician, the Avalon Terrane, by then a defunct volcanic arc, also became increasingly separated from the Gondwanan margin. At about the same time, another large fragment also rifted from a different part of Gondwana to form the new microcontinent of Armorica within the Rheic Ocean.

The Precambrian continental rocks of Armorica were first recognized in northwestern France, and are now known to underlie much of western Europe. The Meguma Terrane of southern mainland Nova Scotia was likely part of Armorica prior to the opening of the Atlantic Ocean. About 390 million years ago, the Meguma Terrane slid into place against the Avalon Terrane from east to west, closing part of the Rheic Ocean. The boundary between the two terranes is marked by a series of faults, collectively known as the Cobequid-Chedabucto Fault System. This system extends through Nova Scotia from Cape Chignecto near Advocate Harbour, past Parrsboro, Truro, and Guysborough and out to sea beyond the town of Canso.

Vertically tilted, thickly bedded Cambrian metasandstone of the Meguma Terrane in a road-cut near Lower Sackville, NS.

The oldest rocks exposed in the Meguma Terrane are quartz-rich sandstones and shales deposited during the Cambrian and Ordovician. This deposition occurred along a passive continental margin of the Rheic Ocean, near what was the African part of Gondwana. The Meguma Terrane continued to evolve in the Silurian as the Rheic Ocean narrowed. As a result, parts of the Meguma Terrane emerged above sea level. This is shown by the early Silurian quartzite of White Rock, NS, which is made from the kind of sand deposits formed on beaches and barrier islands. At other times during the Silurian, volcanoes built up thick, submarine volcanic sequences in rift basins, such as those preserved at Cape Forchu, near Yarmouth, NS. Fossil-rich sediments were deposited in the late Silurian and early Devonian, and are preserved in the Kentville, Torbrook and Bear River areas of Nova Scotia's Annapolis Valley.

Volcanic Arcs and Closing of the Rheic Ocean

In southern New Brunswick, a volcanic arc developed in the early Silurian along the northwestern margin of the Rheic Ocean. This is the Kingston Arc, named after the Kingston Peninsula, where its rocks are best exposed. Late Silurian and early Devonian volcanic rocks on Passamoquoddy Bay, to the west of St. George, may have erupted in a back-arc basin of this arc. A continuation of this Silurian volcanic arc is found within the Miramichi-Bras d'Or Terrane on Cape Breton Island. By the middle Devonian, about 390 million years ago, the Meguma Terrane collided with the accreted terranes of Laurentia, and crust of the Rheic Ocean was completely subducted beneath the Kingston Arc. The impact resulted in another mountain-building episode—the Southern Acadian Orogeny—with further uplift of the Appalachians. At about the same time, the sedimentary rocks of the Meguma Terrane were folded and metamorphosed, and gold-bearing quartz veins developed.

The terranes between Laurentia and Meguma, such as Miramichi-Bras d'Or and Avalon, were caught up in the collision that closed the Rheic Ocean and were pushed laterally along major transform faults. The Southern Acadian Orogeny, like the Northern Acadian Orogeny, also resulted in

Peggys Cove, NS, is situated on a granite batholith—the South Mountain Batholith—of Devonian age.

Devonian red sandstones at St. Andrews, NB, showing "honeycomb" weathering.

the generation, deep in the crust, of granitic magma, which rose into the sedimentary rocks of the Meguma Terrane. The South Mountain Batholith of southern Nova Scotia, the largest in the Appalachians, formed in this way in the middle Devonian, between about 390 and 380 million years ago.

Even after the Rheic Ocean closed, the Meguma Terrane continued to slide westward against the Avalon Terrane along the Cobequid-Chedabucto Fault System. By the end of the Devonian, Laurentia—together with Meguma and the intervening terranes, such as the Avalon and Miramichi-Bras d'Or—formed part of a single continent: Euramerica. This has also been called the "Old Red Sandstone" continent because of the reddish colouring of nonmarine sediments found, especially, in parts of both Europe and North America. Such Devonian red sandstones and conglomerates are preserved, for example, at St. Andrews, NB.

Subduction and collision had thus brought together the Maritime collage, and had closed the Iapetus Ocean and part of the Rheic Ocean. The same processes also thrust up the Appalachian Mountains, then one of the greatest mountain ranges on Earth. It would be tens of millions of years before the open ocean again reached what is today eastern Canada.

Ordovician Life

Ordovician fossils are rare in the Maritimes, but they tell a story of increasing diversity of oceanic life. Although trilobites continued to dominate, echinoids, starfish, crinoids, graptolites and bryozoans all appeared for the first time, and there were many new types of corals and nautiloids. Brachiopods were also common, especially in shallow water covering a sea floor formed of lime-rich sands flanking volcanic islands. In the Cambrian, the dominant brachiopods were primitive forms with unhinged shells, but in the Ordovician, forms with hinged shells were most abundant.

A new group of fossils, the graptolites, became important in the Ordovician. Most species of graptolites were planktonic, floating near the ocean's surface. When they died, they were commonly preserved in mud, which later hardened into shale, siltstone or mudstone. Because graptolites evolved rapidly and spread quickly, they are excellent "index" fossils, their identification commonly giving a precise age to the rocks in which they are found.

A Silurian underwater scene, based on a nonmarine, lagoonal or estuarine assemblage found at Nerepis, NB. All of the fish shown are jawless; the large, blue fish is *Thelodus*; the yellow-brown fish are *Ctenopleura*; and the small green fish are *Cyathaspis*. Between the two *Cyathaspis* specimens, progressing along the bottom, is the horseshoe crab *Bunodella*; the small swimming arthropods are *Ceratiocaris*.

Among the volcanic rocks of the Tetagouche Back-Arc Terrane near Bathurst, NB, possible hydrothermal vent clams (bivalves) have been discovered. Like modern vent clams, they probably derived their energy from the same hot, sulphide-rich waters that produced the Bathurst ore deposits. Paleomagnetic measurements on nearby volcanic rocks suggest that these clams lived in temperate rather than tropical latitudes, and so waters away from the influence of hydrothermal vents were probably cool.

Silurian and Devonian Life

The next major event in the history of life, after the Cambrian "explosion", occurred in the Silurian. This was the emergence of life on land. Although life evolved in the oceans more than 3.5 billion years ago, it was not until about 420 million years ago, in the Silurian, that animals and plants left their

Archaeopteris, a progymnosperm, was among the first trees, and is found in late Devonian and early Carboniferous rocks.

watery home. The most crucial events in plant evolution—the development of stems, leaves and roots—happened during the Devonian. By the end of this period, some plants, such as *Archaeopteris*, rivalled modern trees in size. But more about life on land after we tell the story of life in Silurian and Devonian waters.

During the Silurian, great coral reefs built up in warm, shallow waters in parts of the Maritimes. The presence of these reefs and paleomagnetic data show that the Maritimes was in the tropics, perhaps a few degrees south of the equator. The reefs were also home to other animals, such as brachiopods, clams, snails, crinoids (sea lilies), bryozoans and trilobites. Fossils of these organisms can be seen at Quinn Point, near Jacquet River in northern New Brunswick. Behind the reefs, brachiopods and clams lived on the sea floor; in deeper water, creatures left traces of their activity. Graptolites drifted in the water column, and primitive fish appeared in the Maritimes for the first time. Such fish are preserved in sedimentary rocks at Nerepis, NB.

Arisaig, NS, was home to a diversity of invertebrates during the Silurian, but there were no coral reefs, suggesting that the climate was cooler here than in northern New Brunswick. The rocks at Arisaig form one of the most complete Silurian sections in North America. About 1,400 to 1,500 metres of mainly black to greenish-grey mudstone, shale, and fine-grained sandstone are exposed in cliffs along the Northumberland Strait shore. These rocks were formed in a shallow marine environment, with the sandstones reflecting periodic major storms, producing layers referred to as "tempestites". Because of changing sea levels, two units, both

Specimen of the crinoid *Periechocrinus* from Silurian rocks of the Avalon Terrane at Arisaig, NS.

Lobe-shaped colonies of colonial corals related to *Favosites* in Silurian strata from Quinn Point, near Jacquet River, NB.

A Devonian undersea scene, based on fossils found at Dalhousie, NB. Corals dominate, with both upstanding *Zaphrentis* and low *Favosites*. In the foreground, and in the channel behind, are specimens of the brachiopod *Leptaena*, with smaller brachiopods in the channel. In the distance, crinoids sway in the currents, and a eurypterid, *Pterygotus*, claws outstretched, approaches the channel.

Marine late Silurian sedimentary rocks at Arisaig, NS.

recognizable by their red colour, were deposited under nonmarine conditions, probably in an estuary or river. Some of the sedimentary rocks at Arisaig contain diverse invertebrate fossils. These assemblages are dominated by bottom-dwelling organisms, particularly brachiopods, clams, crinoids and arthropods. The fossils can still be seen in their original life position in the mudstones, but are more commonly found in shell layers concentrated by intense storm waves at the base of tempestites. There are also abundant trace fossils, most produced by worms living within and on the ancient sea floor.

The Maritime provinces remained close to the equator in the Devonian Period. The warmer water was home to a diverse fauna, including corals, brachiopods, fish and sea scorpions (eurypterids), as shown by fossils found near Campbellton, NB. Similar fossils from the Devonian "Old Red Sandstone" in Scotland support the idea that the two areas were part of a single Euramerican landmass. Both marine and freshwater fish have been discovered, including heavily armoured placoderms, jawless fish such as *Yvonaspis*, acanthodians (often called spiny sharks) and true sharks.

The eurypterid *Pterygotus*, which has been found in Silurian rocks at Arisaig, NS and in Devonian rocks at Atholville, NB.

Eurypterids are the largest arthropods known, with some *Pterygotus* growing to more than two metres long. These species likely lived in estuaries or lagoons, sometimes venturing into the shallow ocean. *Pterygotus* was an agile swimmer that used its broad flat tail as a rudder and its large claws for grasping and cutting its prey.

Rocks of the early Devonian of northern and western New Brunswick are nonmarine, indicating that the area was above sea level and that sediment deposition was restricted to river valleys in the rising mountains. Sedimentary rocks containing fossils of plants and fish, exposed between Campbellton and Dalhousie, were probably deposited at the mouth of a tidal river. It was from this area, in the nineteenth century, that J.W. Dawson described *Psilophyton*, one of the earliest known land plants. These rocks remain a source of important fossil discoveries, with finds near Dalhousie Junction, NB—including the earliest evidence of land animals in North America. The fossils include a centipede-like creature called *Eoarthropleura*. Fragments of this arthropleurid (whose giant descendent *Arthropleura*, we will meet in the next chapter) show that the animal may have been 15-20 centimetres long and 2-3 centimetres wide. Also found at the site were fragments of an air-breathing scorpion about 9 centimetres long.

Cross-section through the stem of a Devonian plant similar to *Psilophyton*, Dalhousie Junction, NB.

Devonian early plant fossil, *Psilophyton*, from Dalhousie Junction, NB.

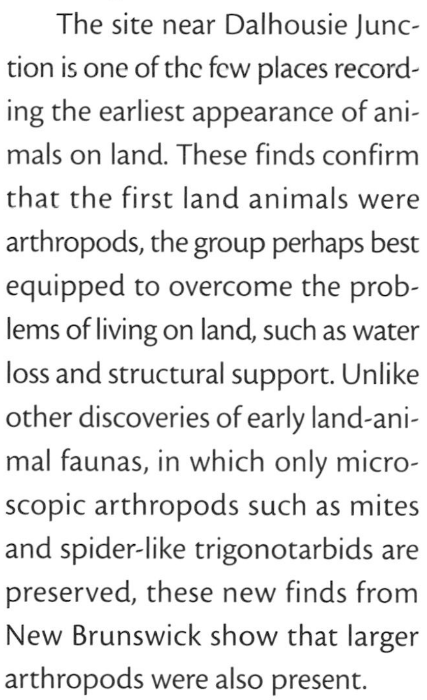

The site near Dalhousie Junction is one of the few places recording the earliest appearance of animals on land. These finds confirm that the first land animals were arthropods, the group perhaps best equipped to overcome the problems of living on land, such as water loss and structural support. Unlike other discoveries of early land-animal faunas, in which only microscopic arthropods such as mites and spider-like trigonotarbids are preserved, these new finds from New Brunswick show that larger arthropods were also present.

In this chapter, we have travelled in time from the earliest Cambrian with its scattered continents, widespread terranes, and barren landscapes, to the late Devonian, with fewer, larger continents, a com-

Psilophyton

pleted Maritime collage, and a landscape populated by trees and arthropods—sparse by modern standards, but far richer than the desolate Cambrian barrens. The scene is almost set for the great Carboniferous coal forests.

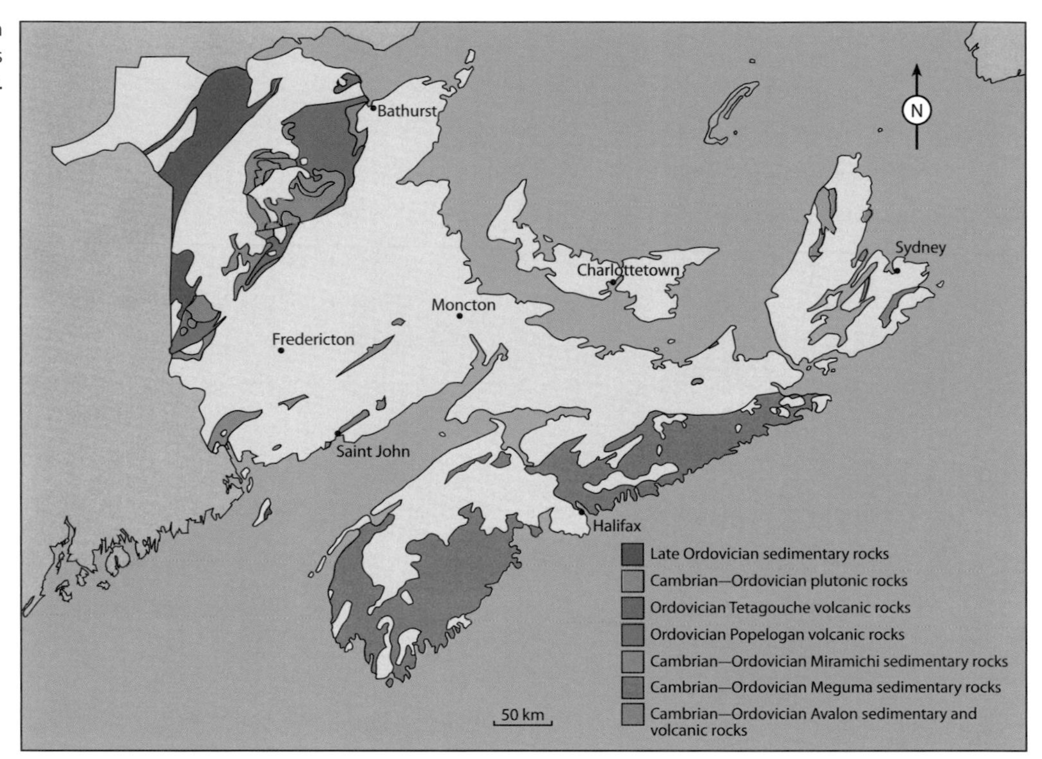

Distribution of Cambrian and Ordovician rocks in the Maritimes.

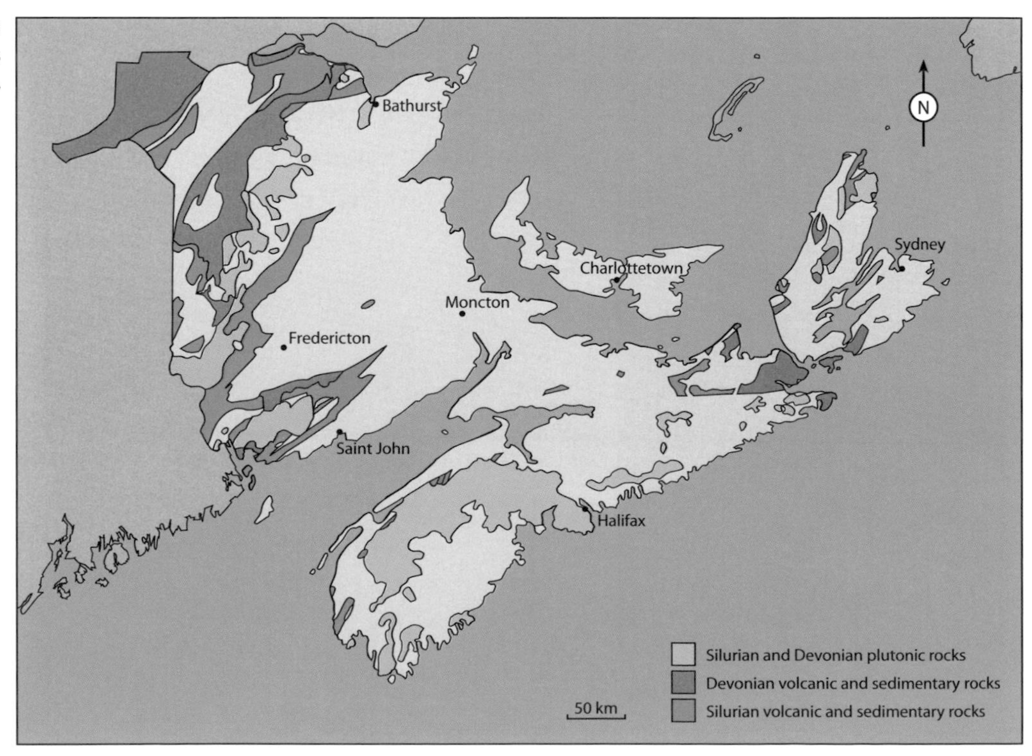

Distribution of Silurian and Devonian rocks in the Maritimes.

Earthquakes and Turbidity Currents

The Grand Banks Earthquake

On a mid-November afternoon in 1929, the year of the great stock market crash, the earth shook in Atlantic Canada. The effects were felt in many parts of Nova Scotia. According to one report from Canso, "It seemed as though a large steamer had bumped into the wharf and the bumping continued as though the engine and propeller were still going". In Ingonish, cups were upset, windows were broken and a concrete door sill was jarred out of alignment. In Halifax and Lunenburg, houses vibrated. Chimneys toppled in Sydney. A fire bell was set off in Bridgewater, and several parked cars started to move. Even as far away as Yarmouth, small items were dislodged, but people walking on the street felt nothing. These disturbing events were triggered by an earthquake of magnitude 7.2 on the Richter Scale, with its epicentre at the mouth of the Laurentian Channel. This earthquake came to be known as the Grand Banks Earthquake.

Earthquakes of such magnitude centred beneath the ocean tend to set off huge waves of water. These sea waves are known by the Japanese word *tsunami* (meaning "harbour wave"). They are also called "tidal waves", but this is a misnomer, since they have nothing to do with tides. The waves produced by the Grand Banks Earthquake were no exception. The resulting tsunami caused much destruction as well as the death of 28 people, almost all in Newfoundland's Burin Peninsula. Ironically, with all the talk of the imminent "Big One" on the west coast, these east coast deaths are the only earthquake fatalities recorded in Canada. In addition to the loss of life, the waves wrecked boats, wharves, and houses close to the shore.

The Discovery of Turbidity Currents

A third effect of the Grand Banks earthquake, after the shaking of the ground and the tsunami, was the most puzzling. It involved the snapping of numerous trans-Atlantic telephone cables that had been laid under the ocean south of Newfoundland. Those cables near the epicentre of the earthquake broke during the quake, which is not surprising. However, what was surprising was the progressively longer time delay of cable breaks farther from the epicentre. The most distant cable, 600 kilometres south, broke 13 hours after the earthquake. Scientists at the time were baffled by this progressive sequence of breaks.

Twenty-five years later, south of the earthquake site, a layer of sand was dis-

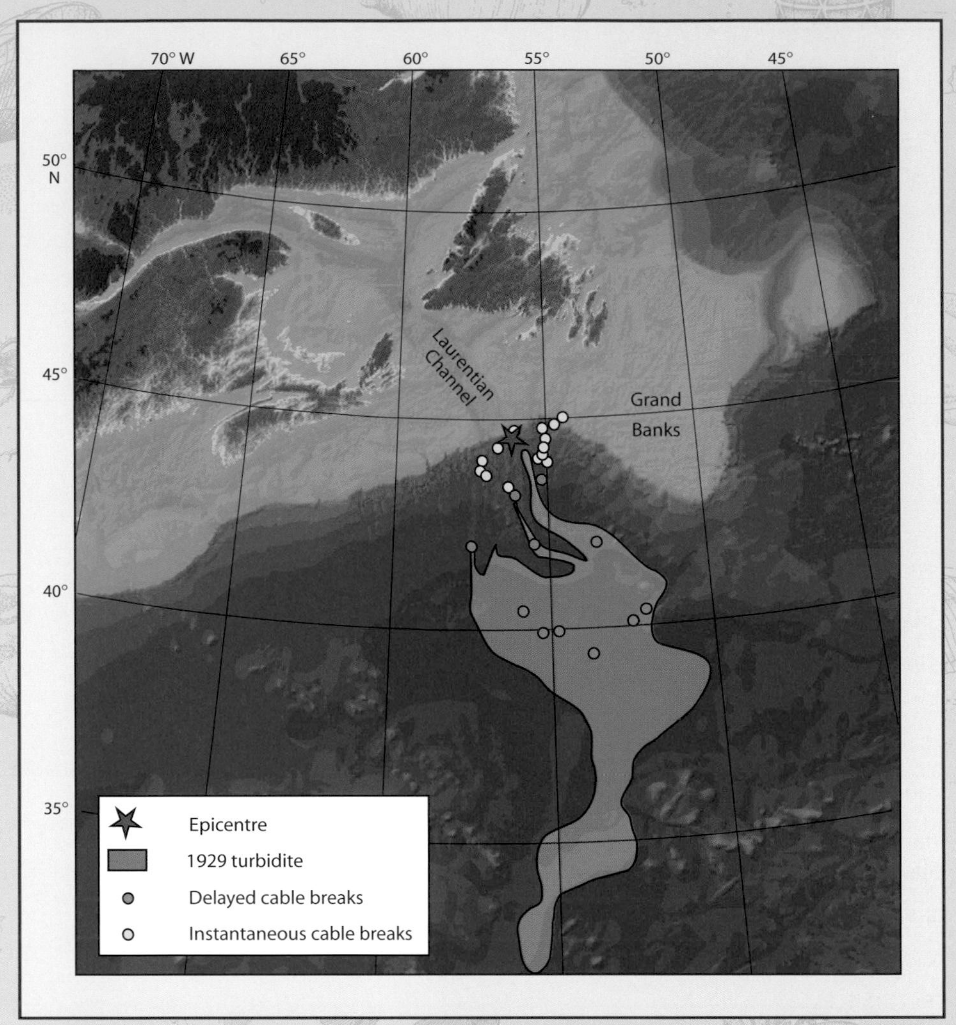

The extent of the turbidite from the Grand Banks earthquake and the timing of cable breaks.

covered on the sea floor, covering an area the size of the province of Quebec. And evidence of vast underwater landslides was found near the epicentre. These finds led to a theory that explained not just the cable breaks, but a whole class of sedimentary rocks. According to this theory, catastrophic underwater currents, called "turbidity currents", carry large amounts of sediment from the continental slope to the deep sea floor. The turbidity current triggered by the 1929 Grand Banks earthquake travelled at speeds of up to 65 kilometres per hour down an undersea valley 25 kilometres wide and 300 metres deep. The volume of sand and mud that it transported would be sufficient to completely fill Nova Scotia's Annapolis Valley. Sedimentary rocks derived from such sediment slides are called "turbidites". They are common, for example, in the strata of the Meguma Terrane around Halifax and on mainland Nova Scotia's southern and eastern shores, as well as in the Miramichi-Bras d'Or Terrane of New Brunswick.

Other Earthquakes that Affected the Maritimes

The Grand Banks Earthquake is perhaps the most famous earthquake that has affected the Maritime Provinces, but it is neither the earliest recorded nor the largest known. In 1764, the *Halifax Gazette* gave the first written account of an earthquake in the region, felt around Saint John, NB. Since then, over a hundred relatively mild earthquakes have been reported in the Maritimes, mostly in New Brunswick, clustered in the Passamaquoddy Bay area and in the Miramichi Highlands.

The largest modern onshore earthquake in the Maritimes occurred on 9th January 1982 in the Miramichi Highlands, about 80 kilometres west of the city of Miramichi. The main shock, of magnitude 5.7 on the Richter Scale, was felt just before nine o'clock in the morning, followed by aftershocks of magnitudes 5.1 (later the same day) and 5.4 (two days later). Although ground motion was felt over most of the province, damage was limited to hairline fractures in the walls of a few buildings in Miramichi, Bathurst and Perth-Andover. No fault trace was found that could account for the earthquake, but this is not surprising as the area is heavily wooded and has a thick soil.

As mentioned in Chapter 1, most earthquakes happen along tectonic plate boundaries, but the Maritime Provinces are presently well within the large North American Plate. Why, then, are there any earthquakes at all in our region, even rare and relatively weak ones? The reason is that, although most tectonic action is indeed at plate boundaries, the plates themselves are in motion. This causes a build-up of stress within the North American Plate as it drifts slowly westward over the Pacific plate, and it is this stress that is released occasionally by mid-plate earthquakes in the Maritimes and elsewhere.

This house from Port au Bras, Newfoundland, was removed from its foundations by the tsunami that followed the 1929 Grand Banks Earthquake. Here, it is tied up for safekeeping to the Lunenburg schooner *Marian Belle Wolfe*, anchored in Little Burin Harbour. The photograph was taken by Father James Anthony Miller on19 November 1929, the day following the tsunami.

The pond behind the beach at Lords Cove, Newfoundland, is strewn with debris and buildings as a result of the tsunami that followed the Grand Banks Earthquake. In the gabled house to the left, a mother and two children on the first floor were drowned, but a third child, in bed on the second floor, slept through the disaster and survived. The photograph was taken by H.M. Mosdell, the local Member of the House of Assembly at the time.

Granites

Intrusive Igneous Rocks

Igneous rocks form from the cooling of magma, a hot (600 to 1,200°C) mixture of liquid rock, solid minerals and gas. When magma reaches the Earth's surface, it quickly cools and crystallizes, within a few seconds to a few months, to become pyroclastic material or lava flows. Because of this rapid cooling, new rock formed from magma consists of small, generally microscopic crystals. Examples of such "extrusive" igneous rocks are the dark-coloured early Jurassic volcanic rocks (basalts) of the North Mountain in Nova Scotia and the light-coloured Devonian volcanic rocks (rhyolites) at Campbellton, NB.

Not all magma reaches the Earth's surface, however. Much of it remains underground where, insulated by the surrounding rocks, it cools and crystallizes slowly over thousands or even millions of years. This process results in coarser-grained igneous rocks, with crystals clearly visible to the naked eye. Such igneous rocks formed underground are "intrusive". Large bodies of intrusive rocks are called "plutons", from the Greco-Roman god Pluto, who ruled over the underworld. Granites are the most common plutonic rocks on Earth.

Uplift and erosion have exposed many granite plutons at the surface today. An example is the South Mountain Batholith ("batholith" is the term for a large type of pluton), which forms the backbone of southwestern mainland Nova Scotia. Peggys Cove lighthouse, one of Nova Scotia's most famous landmarks, stands dramatically on glacier-worn granite of the South Mountain Batholith. Plutons exposed in New Brunswick include the Pabineau Granite, near Bathurst.

Scene at Peggys Cove, NS, showing the lighthouse and glacially-polished granite of the South Mountain Batholith. Notice the cracks (or "joints") in the granite.

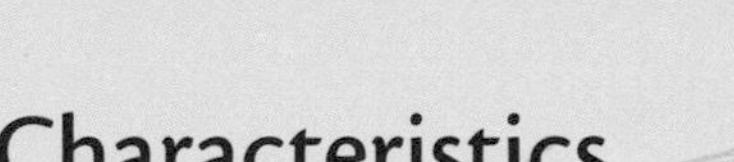

Characteristics of Granites

The main minerals in granite are quartz and two types of feldspar—alkali feldspar and plagioclase. Quartz has a glassy appearance, alkali feldspar may be white or pink, and plagioclase tends to be white or buff. Small amounts of other minerals may also be present, especially shiny flakes of mica (black biotite and light-coloured muscovite), and dark green or black hornblende. Granitic rocks are normally pale, but they vary from white to grey, buff, pink, or red, depending on the minerals, and proportions of minerals, present.

Crystals in granites are usually a few millimetres to a few centimetres across, but finer-grained granitic rocks, called "aplites", form as a result of rapid crystallization. Very coarse-grained granitic rocks, with crystal sizes ranging from centimetres to metres, called "pegmatites" form during slow crystallization in the presence of water vapour. Many granites have two distinct sizes of crystals. The large crystals, called "phenocrysts", formed in an episode of early slow cooling in the magma. The smaller crystals, called "matrix" or "groundmass", crystallized more quickly.

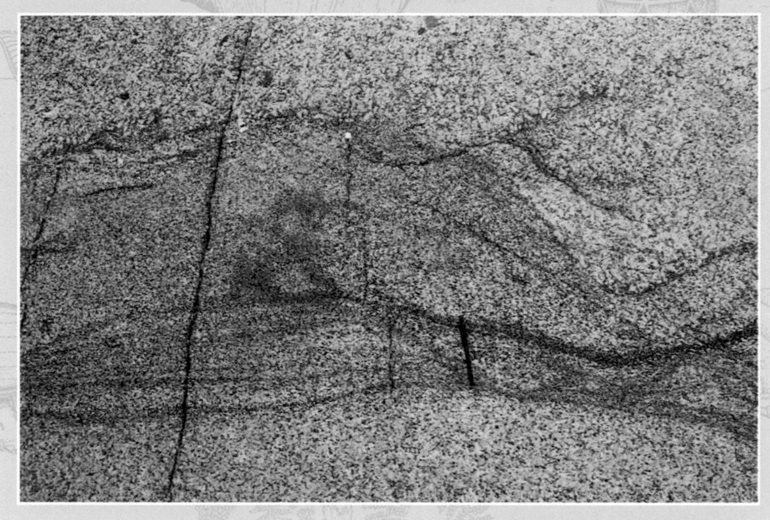

Schlieren, or streaks of dark biotite, in granite at Prospect, NS.

As a granite magma cools to the temperature of its surroundings, it may develop a number of structural features. The magma may flow during crystallization, sometimes forming distinct layers or bands of dark minerals. Such bands are called "schlieren". Some granites are "foliated", with aligned, elongate minerals such as feldspar or mica, produced by stresses during or after crystallization. Such rocks are called "granite gneisses". Finally, because of cooling and contraction, most granite bodies have a regular system of shrinkage cracks known as "joints".

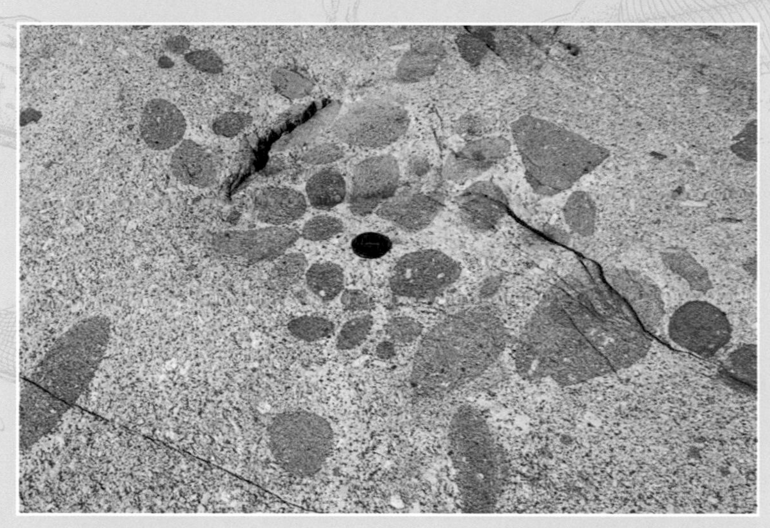

Cluster of dark granite enclaves in granite near Sambro, NS.

How Granites Form

Most granite plutons occur in places where two continents, or an ocean and a continent, have collided to form mountain ranges. Such collisions commonly produce granite magmas by partial melting of deep crustal (or possibly even mantle) rocks at temperatures of 600 to 900°C. The molten part of the source rock eventually separates from the solid part, much as water separates from a wet sponge when squeezed. The magma, being less dense and less viscous than the surrounding rocks, moves upward. It may rise tens of kilometres by flowing in vertical fractures, by ascending as diapiric blobs (as in a lava lamp), and by "stoping". Stoping is the process by which solid blocks of the roof of a pluton break away and exchange places with the magma.

Large xenolith of metasedimentary rocks of the Meguma Terrane, in granite at Portuguese Cove, NS.

Enclaves

As granite magmas rise through the crust, they may pick up and carry fragments (inclusions or "enclaves") of the surrounding solid rocks. Some enclaves are previously crystallized pieces of the granite itself (called "autoliths"), with slightly different minerals or textures that make them stand out. Especially during stoping, however, the magma can pick up enclaves foreign to the granite (called "xenoliths", from the Greek *xenos*, meaning foreign or strange, and *lithos*, meaning rock). Xenoliths are easy to recognize because they differ greatly in mineral composition, colour and texture from the enclosing granite. They are usually less than a metre in size, but some may be hundreds of metres across.

Most xenoliths are denser than granite magma, so why don't they all sink and collect at the bottom of the pluton? The lack of such collections of xenoliths (called "elephants' graveyards") has been a puzzle for geologists. But one large xenolith from the South Mountain Batholith at Portuguese Cove, NS may offer some clues. This particular xenolith is twenty to thirty metres across and consists of well-bedded metasedimentary rocks, stoped from the surrounding Meguma Terrane. Because of its large size, this xenolith should not have remained suspended in the magma. The hundreds of granite veins running through it show clearly that it was breaking apart. This process of fragmentation may have involved small explosions, much as ice cubes explode when dropped in a drink. Perhaps the smaller blocks created by this break-up of the larger xenolith stayed suspended because of convection currents in the magma. The Portuguese Cove example suggests that most xenoliths disintegrate before they reach the bottom of the pluton, explaining the lack of "elephants' graveyards".

Cluster of dark metamorphic xenoliths associated with the large xenolith at Portuguese Cove, NS.

CHAPTER 6
Basins and Ranges

A Permian landscape, with a group of *Dimetrodon* on the prowl among cycad bushes (foreground, left and right) and horsetail tree (middle distance to left).

Ancient Earthquakes

In 1981, geologist Guy Plint, then at the University of New Brunswick, was studying Carboniferous rocks, at Tynemouth Creek near Gardner Creek, NB, a few kilometres north of the Cobequid-Chedabucto Fault System. As we saw in the previous chapter, this fault system marks a major ancient plate boundary between the Meguma Terrane of the microcontinent of Armorica and the Avalon Terrane, which was attached to Euramerica by the Carboniferous. Within a sandstone layer at Tynemouth Creek, Plint noticed a curious, narrow fissure that contained tilted sandstone blocks. At first he thought that, far back in time, the rocks had been broken up by movement along a fault.

However, closer inspection showed some puzzling features. The strata in the cliffs above the fissure showed no signs of fracturing and rested upon the broken blocks. This

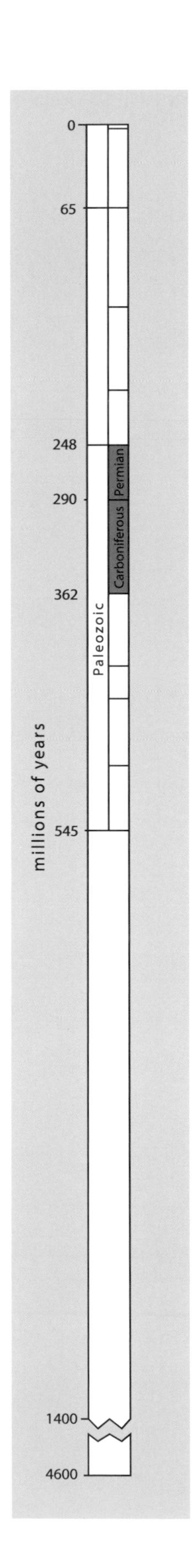

suggested to Plint that the event that created the fissure had taken place during the Carboniferous, between the accumulation of the sand layer containing the fissure and the accumulation of the overlying sediments. In an overhanging lip above the fissure, he also observed the fossilized remains of roots, as well as the trackway of *Arthropleura*, a gigantic millipede-like arthropod. This scene reminded Plint of photographs he had seen of fissures near the San Andreas Fault in California, where strong surface earthquake waves had torn open cracks in the ground. The photos showed blocks of earth and soil that had caved into the fissures (along with cars and houses), while the roots of grass and bushes had strengthened the surface soil, forming an overhang.

With these new insights, Plint discovered more evidence of disturbance at Tynemouth Creek, including deformed and fissured mudstone. This suggested that soft, watery muds had been disturbed and even injected into stronger sediment layers. Indeed, Plint had documented a rare feature in the geological record: direct evidence of a powerful earthquake far back in time.

The Tynemouth Creek earthquake fissures tell us that there was active movement along the Cobequid-Chedabucto Fault System (much like the modern San Andreas Fault), as plates came together in the final stages of the assembly of the supercontinent Pangea.

Birth of the Maritimes Basin

After the initial collision of Armorica and Euramerica in the Silurian and Devonian, these continents continued to move in the succeeding Carboniferous Period. The Armorican Meguma Terrane inched westward past the Avalon Terrane on the margin of Euramerica along the Cobequid-Chedabucto Fault System, resulting in powerful earthquakes. This fault system has had a profound influence on the geology and modern landscape of the Maritimes, especially that of Nova Scotia.

The closure of the Iapetus and Rheic oceans—and the consequent plate collisions—resulted in mountain ranges along what is now the eastern coast of North America, from Alabama in the southwest to Newfoundland in the northeast. The modern Appalachians are all that remain of these ancient mountains. Since there was no Atlantic Ocean at the time, the mountains continued through Greenland to northwestern Europe. Areas within the mountains slipped and subsided along many faults associated with the Cobequid-Chedabucto and other fault systems, producing small, deep basins. ("Basin" is the term used by geologists to describe a part of the Earth's crust that sinks or "subsides" to allow great thicknesses of sediment—in some cases up to about 20 kilometres—to accumulate.)

Early Carboniferous river channel deposits, shown by the bulge in the rock face at lower right. Victoria Park, Truro, NS.

This mosaic of interconnected mountains and basins, collectively called the Maritimes Basin, dominated the landscape of the Maritimes from the late Devonian to the Permian. The Maritimes Basin extended from central Nova Scotia northward across the Gulf of St. Lawrence to the present-day shore of the Gaspé Peninsula, and from western New Brunswick to Newfoundland. Rivers, streams and mudflows transported gravel, sand and mud into the basins, gradually building up thick layers of sediment. In places, volcanoes erupted along lines of weakness marked by faults.

From about 365 to 340 million years ago, in the latest Devonian to early Carboniferous, in the lowlands within the Maritimes Basin, there were extensive river systems. There were also deep lakes bordered by deltas (alluvial fans) formed from sands and gravels washed down in mountain streams. Microscopic algae thrived in the lakes, and their remains formed thick layers of organic ooze on the lake bottoms. After burial and over time, this ooze produced "oil shales", such as the Albert shales of southern New Brunswick. These deposits are so rich that petroleum products derived from them form a small oil and natural gas field near Moncton, NB, which produced oil and natural gas for 80 years. A new phase of exploration for natural gas is now underway.

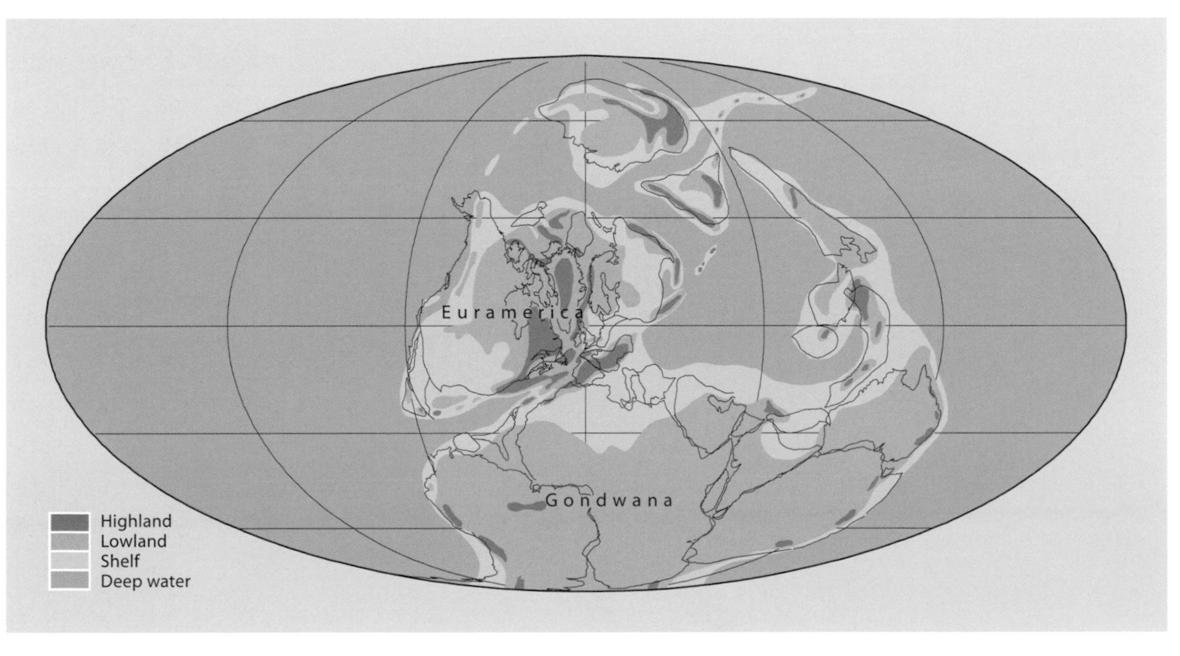

Global paleogeography of the early Carboniferous, about 340 million years ago.

Regional paleogeography in the early Carboniferous, about 355 million years ago.

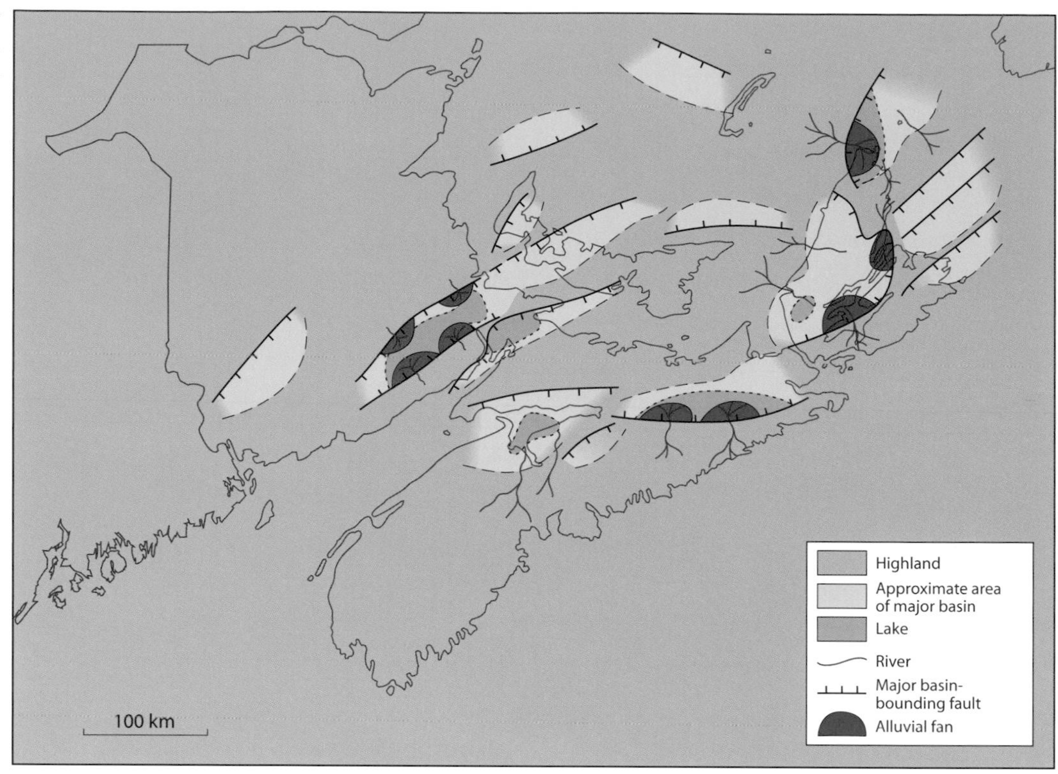

The oil shales are also rich in fossils of some of the earliest bony fish, called palaeoniscoids. These were small fish, but there were also larger, more ferocious-looking fish in the early Carboniferous lakes of the region. We know this from the recent discovery of a 20-centimetre-long jawbone of a spike-toothed crossopterygian fish at Blue Beach, near Hantsport, NS. Another crossopterygian fish, of similar age, was found north of Moncton, NB. Crossopterygians are related to the lobe-finned fish that gave rise to amphibians, the first vertebrates to walk on land—an important step, literally, in the evolution of life.

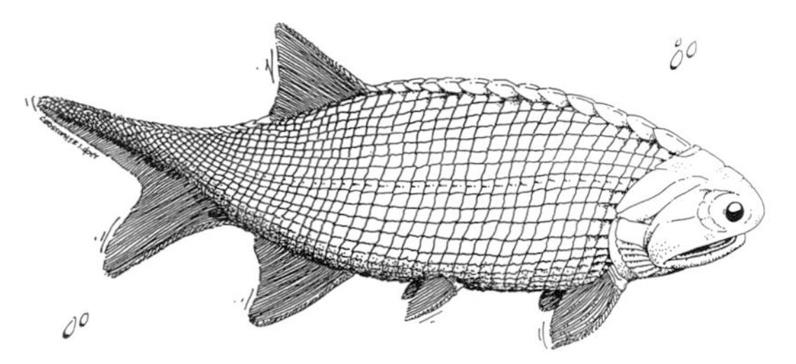

The palaeoniscoid fish *Elonichthys*, found in early Carboniferous rocks at Albert Mines, NB.

Some of the earliest footprints of ancient amphibians are found in sedimentary rocks formed along ancient lake shorelines, now exposed at the mouth of the Avon River at Blue Beach. One remarkable trackway, discovered by Sir William Logan, first Director of the Geological Survey of Canada, records the activity of an unidentified animal larger than a crocodile. The discovery was made in 1841, when a slab of building stone quarried at Horton Bluff, near Blue Beach, caught Logan's eye. On the slab were the oldest fossil footprints discovered up to that time—the first to be recognized from rocks older than Permian. Logan's report of this find to the Geological Society of London was largely ignored until, some years later, Sir William Dawson highlighted the discovery in his pioneering book,

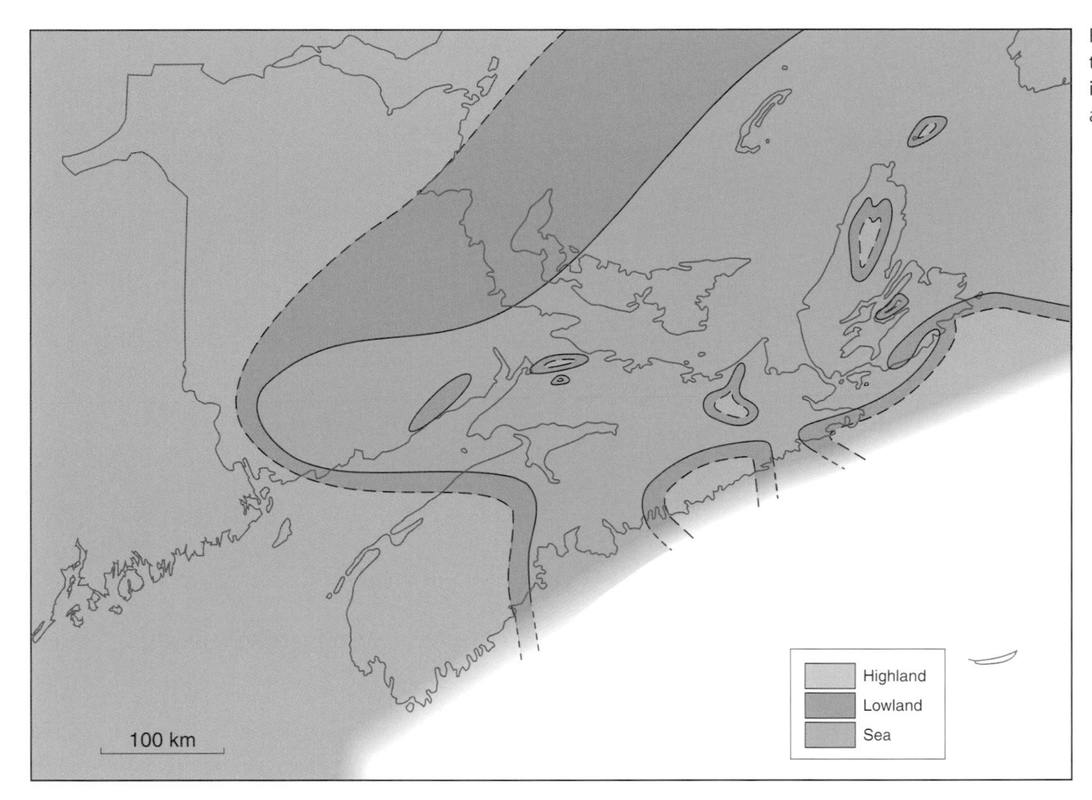

Regional paleogeography at the time of the Windsor Sea, in the early Carboniferous, about 335 million years ago.

Acadian Geology. Plant fossils, such as those of early clubmosses, are also common in some of these earliest Carboniferous rocks.

The Maritime Tropical Sea

A great transformation of the Maritime landscape took place about 340 million years ago. The climate was becoming drier as the continents came together to form the supercontinent Pangea. The region that was to become the Maritimes lay within Pangea, at or near the equator. Surrounding Pangea was a huge ocean, the Panthalassic. Sea water began to invade the deeper, subsiding depressions in the Maritimes Basin. Eventually the sea covered all but the mountainous highlands in Nova Scotia and Prince Edward Island, and extended as far as central New Brunswick. This ancient body of salt water is called the Windsor Sea, after the Windsor area of Nova Scotia, which is underlain by sedimentary rocks formed on the sea's floor. Deposits of the Windsor Sea also occur near Crystal Cliffs, NS; on Cape Breton Island; and in the Sussex and Hillsborough areas of southern New Brunswick.

Sea level rose and fell periodically during the 15 million years that the Windsor Sea existed in the Maritimes Basin. We know this from the repeating record of carbonates, evaporites (such as gypsum, rock salt and potash),

An early Carboniferous amphibian trackway from Blue Beach, near Hantsport, NS. Ripple marks, formed on the ancient shallow lake bed, were trampled by the amphibian.

Gypsum, which forms these cliffs near Windsor, NS, was deposited in the Windsor Sea.

and "red beds" (such as shales, sandstones and conglomerates). These cyclical changes reflect changes in climate between more and less arid intervals, which in turn may have been caused by the periodic expansion and shrinkage of a huge ice cap over the South Pole, far to the south of the tropical Maritimes. The ice cap's fluctuations drastically lowered and raised global sea levels, as vast amounts of water were respectively frozen and thawed.

In many areas, the climate was at its most arid when sea level fell and water retreated from the continent. As sea water evaporated under the intense tropical sun, it became highly concentrated in minerals, precipitating first gypsum (calcium sulphate), then rock salt (sodium chloride) and finally potash (potassium chloride) on the bottom of shallowing seas. Today these evaporites, which are mined in southern New Brunswick and northern Nova Scotia, extend northward under most of Prince Edward Island and the eastern Gulf of St. Lawrence (see Chapter 8). The next time you sprinkle salt on your French fries, pause to consider that it may have precipitated from the Windsor Sea over 300 million years ago!

Rock salt, gypsum and potash are softer and more buoyant than most other sedimentary materials. These soft rocks can be easily deformed by the weight of younger sediments deposited on top of them or by movement along faults. In places, deposits of the Windsor Sea have become deformed into complex, column-like masses (called "diapirs") that can be several kilometres wide and high. Where thick evaporite layers are present (especially to the south of the Magdalen Islands in the Gulf of St. Lawrence and along the coast of western Cape Breton) numerous diapirs have been detected by remote sensing techniques. The upward movement of diapirs has disturbed the surrounding rocks, producing some spectacular cliff scenery, as at Crystal Cliffs, near Antigonish, NS, and in the Mabou area of Cape Breton Island, NS. The salt mine at Pugwash, NS, and the potash mine at Sussex, NB, both exploit diapirs. Any faults or fractures in the diapir tend to be "healed" by the movement of salt, allowing the salt miners to open large tun-

A trio of ladyslippers. These plants favour soils overlying gypsum deposits.

Gypsum (the lighter-coloured rock) associated with a diapir that has pushed upward through younger rocks at Mabou Mines, NS.

nels deep underground, with ceilings as high as four-storey buildings, without serious danger of roof collapse.

Evaporites are also easily dissolved by ground water, with the ground collapsing or sinking into depressions thus formed. Such "sinkholes", which may now be lakes, are a feature of a landscape underlain by salt and other easily dissolved rocks, such as limestones. This is called a "karst" landscape. Sediments and fossils from later times, such as the Cretaceous (about 140 to 65 million years ago) and the Quaternary (about 1.8 million to a few thousand years ago), have been found in some Maritime sinkholes.

Life in the Windsor Sea was similar to that in modern shallow tropical seas. In places, the sea floor shallowed into limey banks built up by colonies of corals, sponges and lacy bryozoa. Banks developed mainly where oceanic waters lapped over bedrock promontories or against islands that may have provided a firm foundation for the attachment and growth of marine life. Such banks are preserved near Gays River, NS, where they host important zinc and lead deposits. The warm waters above these banks were home to brachiopods, snails, bivalves and a variety of coiled and conical nautiloids. Shelly limestones called "coquinas" are formed almost entirely of these shellfish. Surprisingly, the record of fish from the Windsor Sea is sparse.

A pond filling a sinkhole in Windsor gypsum deposits, beside the Trans Canada Highway, near Baddeck, NS.

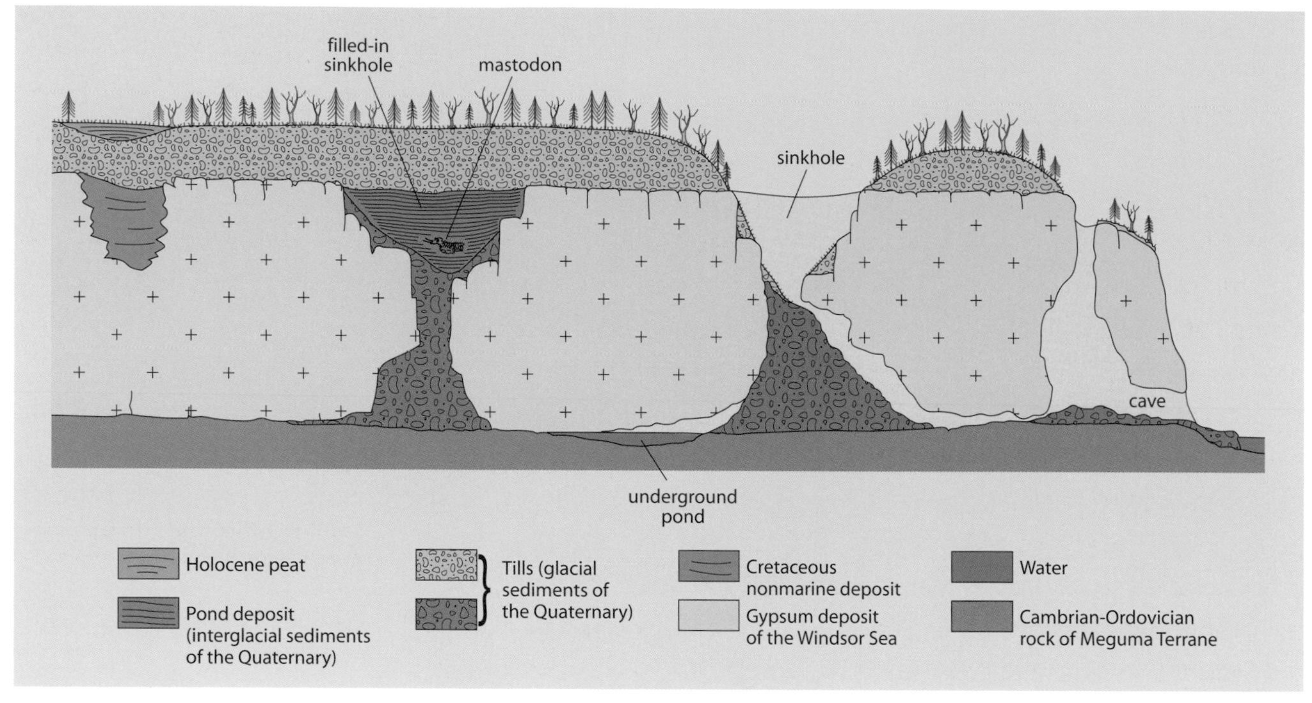

Cross-section of a karst landscape formed in Windsor Sea gypsum deposits, showing Cretaceous, Ice Age and modern sinkholes, as well as modern caves.

After the Windsor Sea receded, red sandstones, shales and conglomerates accumulated beneath the tropical sun, where once the sea had been. The rusty colour of these "red beds" is due to the oxidation of iron during alternating wet and dry conditions, showing that these sediments were deposited on land, rather than in the sea. The climate remained hot, with long periods of drought, and was too harsh for lush vegetation. From time to time, salt lakes (like the modern Great Salt Lake in Utah) dried up and left behind thin deposits of gypsum. Deeper and more hospitable lakes developed along the line of the Cobequid-Chedabucto Fault System in northern Nova Scotia. They supported shrimp-like crustaceans, horseshoe crabs, and fish, including acanthodians and sharks. Amphibians left their traces along the lake shores, as shown by fossil footprints in the rocks of West Bay, near Parrsboro, NS. As the global dance of the continents slowly changed the climate, rainfall increased, and the harsh, arid landscape began to turn green.

Amphibian footprints, known as *Hylopus*, from the late Carboniferous rocks of West Bay, NS. This view is of the underside of the bed, shown by the footprints being raised rather than depressed. The many small raised bumps represent small pits in the original sediment, and probably echo a late Carboniferous rain shower.

The Coal Age

Late Carboniferous coal seam exposed on the beach at Victoria Mines, near New Victoria, NS.

The late Carboniferous began about 325 million years ago and lasted for about 30 million years. It is known as the "Coal Age" because many of the extensive northern hemisphere coal deposits, which have been mined widely and which provided the power for the Industrial Revolution and beyond, were formed at this time. The Maritime Provinces straddled the equator, but were inching their way northward as the continents continued to assemble. Far to the south, the waxings and wanings of the polar ice cap continued to raise and lower global sea levels, just as they had during the time of the Windsor Sea.

This region, now completely emerged from the Windsor Sea, was crossed by northeasterly flowing rivers. The largest of these may have originated from distant parts of the Appalachian Mountains in what is now the eastern United States. The Appalachians were being thrust upward by the slow, relentless collision of tectonic plates, a collision that had begun over 100 million years earlier in the Ordovician and was now approaching its final stages. Between the mountain ranges, extensive rainforest swamps and bogs flourished across the tropical lowlands of the Maritimes Basin as the climate grew more humid. It was in these swamps and bogs that coal formed.

Coal is peat that has been compressed and converted to rock. Peat is a soil of mainly plant material that forms in wetlands or "mires". Swamps,

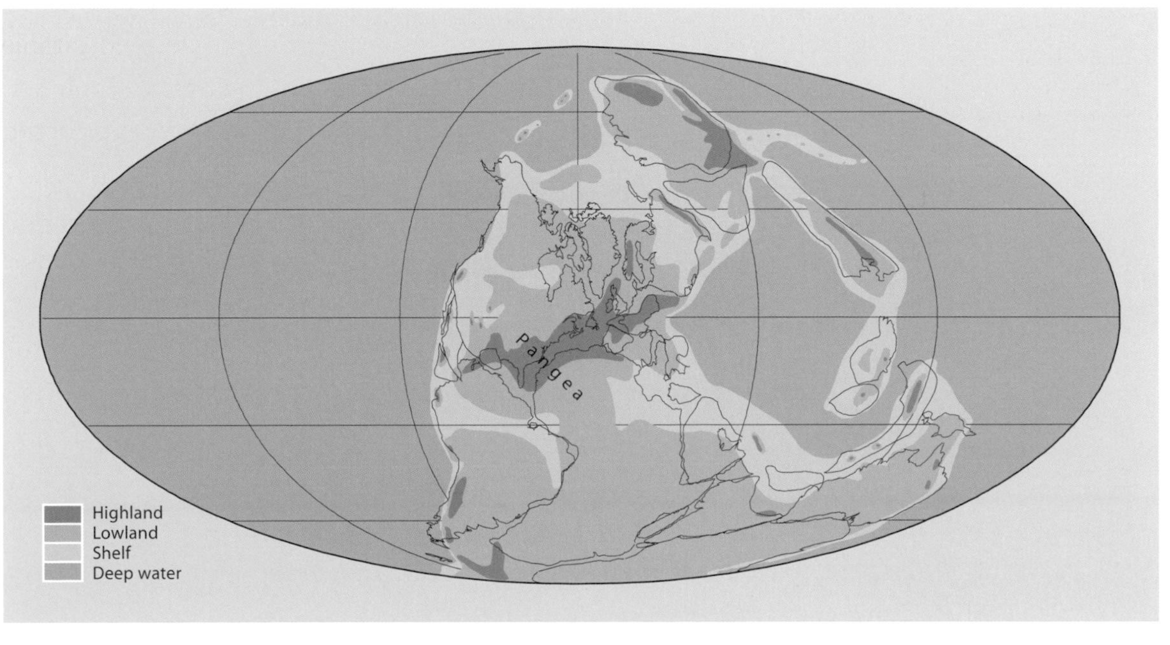

Global paleogeography of the late Carboniferous, about 305 million years ago.

Regional paleogeography in the late Carboniferous, about 310 million years ago.

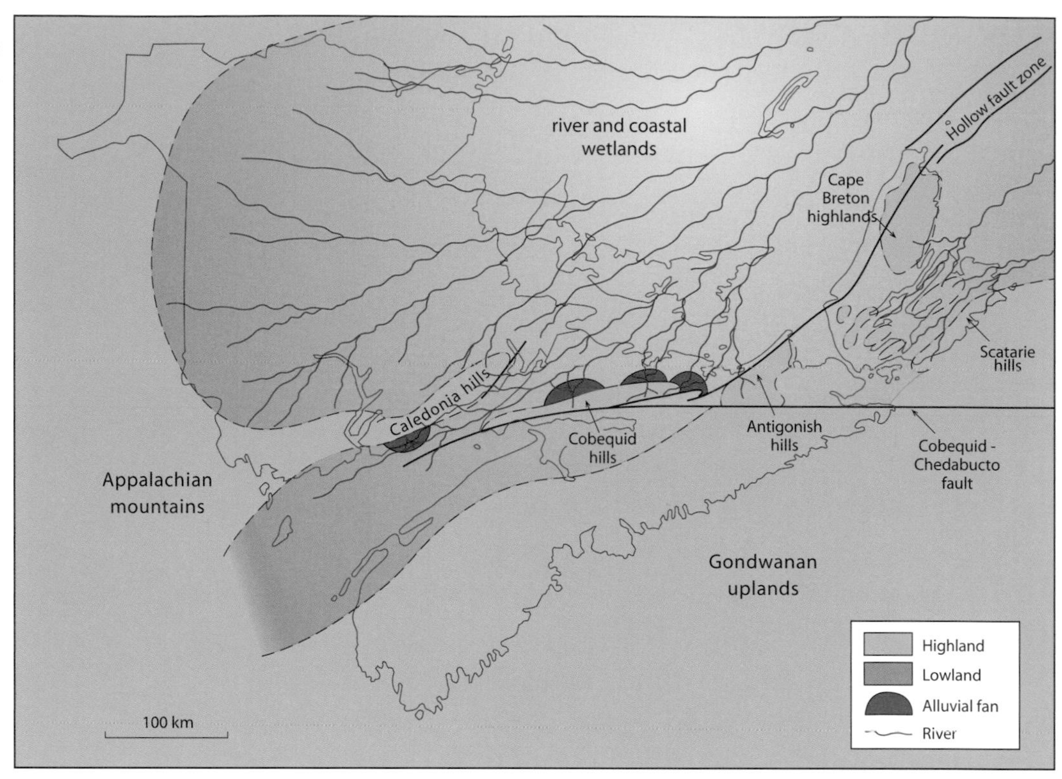

marshes, fens and bogs are all types of mire. Peat accumulates slowly, growing only four millimetres a year in the modern tropics. As peat becomes buried beneath other sediments, it is compressed, and large amounts of water are squeezed out. Over millions of years, the peat hardens, first to lignite and, eventually, to coal, which can be one-tenth the thickness of the original peat. A coal seam one metre thick was originally 5-10 metres of peat that took perhaps 2,500 years to accumulate. The Minto Seam of New Brunswick and many of the coal beds of Joggins, NS, each represent about 1,000 years of peat accumulation. But this is only a short time when compared to the Foord Seam of the Pictou Coalfield, NS—at over 13 metres thick, it may have taken over 30,000 years to accumulate.

When coal is examined in a microscope, details of the original plants are evident. The historical events of the primeval forest swamp—wildfire, flood and the succession of plants—are all recorded within the layers of coal. Coal miners can blame the charcoal from ancient wildfires for the sooty black coal dust on their hands and faces.

Coal was not the only rock formed in the late Carboniferous world. In the Maritimes, areas such as the Sydney, Cumberland and Minto basins were subsiding and being filled by repeated sequences of coal seams, floodplain mudstones, lake or marine limestones and river bed sandstones.

These repeating sequences, known as "cyclothems", record climate and sea level changes and migrating river systems. They can be seen today in the splendid sea cliff exposures of our Maritime coal basins—in places such as Joggins, NS; along the sea coast of Cape Breton Island between Port Hood and Cheticamp; from Boularderie Island to Port Morien in the Sydney Coalfield of Cape Breton Island; and in the Clifton and Cape Enrage areas of New Brunswick.

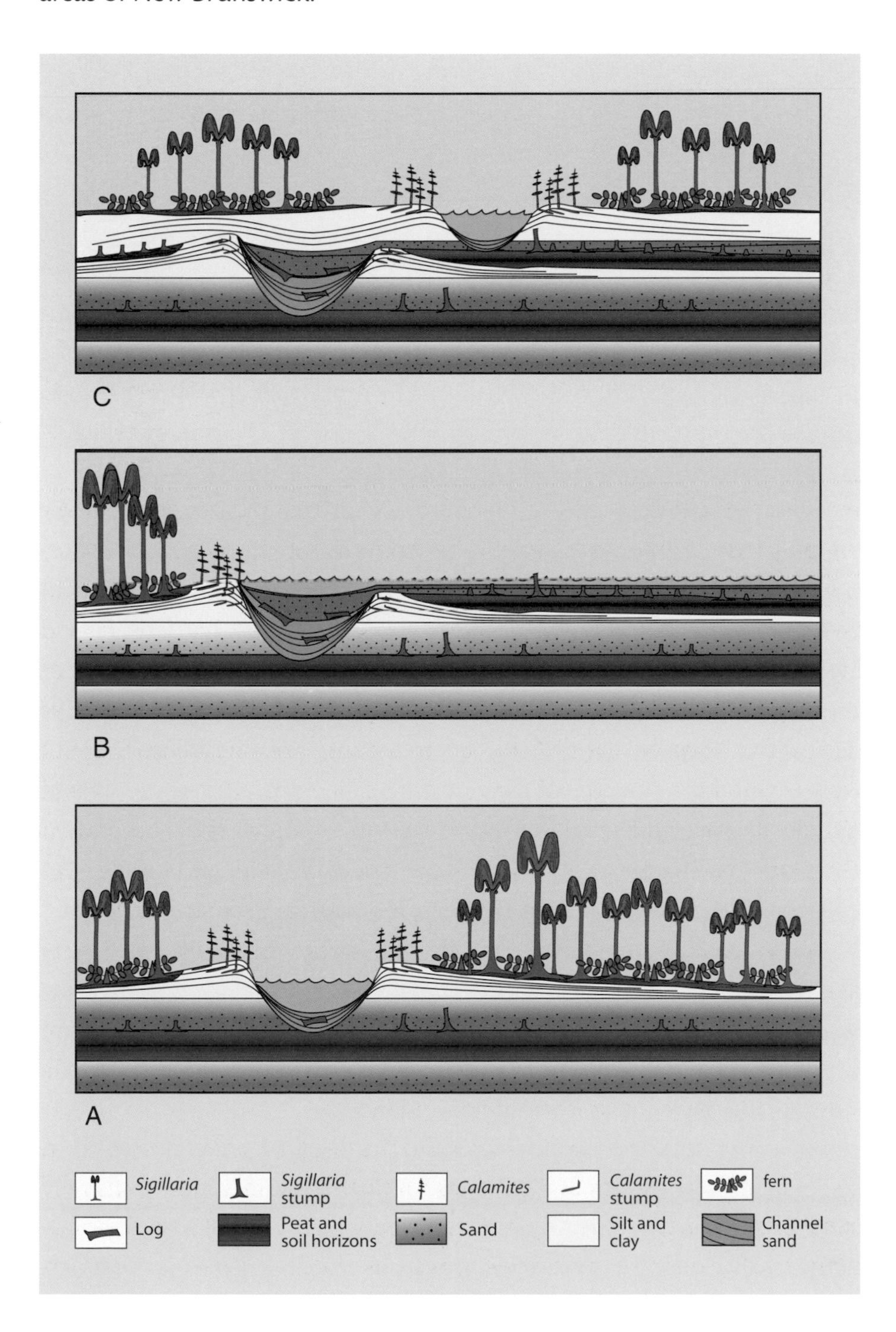

Cycles in a Coal Age forest. A) This sketch shows a river cutting a channel in older sediments, and depositing sand and gravel in its channel. *Calamites* is growing along the banks, beyond which is a flood plain supporting a forest of *Sigillaria* and ferns. Organic debris from the forest is building up a layer of peat that may eventually form coal. B) The river breaks out of its channel during a flood. Sand is deposited by the flood waters, choking the forest, covering the peat and filling the old channel. Only the stumps of the trees are preserved, protected by the new deposit of "sheet" sand. C) A new river establishes a channel, and a new forest is growing on the flood plain, thus beginning a new cycle. Through time, this process builds up repeating layers of peat, sand, silt and clay to produce cyclothems. Marine or brackish water invasions commonly add an extra environment and type of sediment layer not shown here.

Cliffs of late Carboniferous sedimentary rocks along the coast near Margaree Harbour, NS.

At the margins of the basins, adjacent to highland ranges, alluvial fans of coarser clastic sediments accumulated. Here, upland streams were checked in their progress and so deposited their coarse sand and gravel load. What caused the changes in climate and sea level that resulted in these cycles of sediment?

One possibility is cyclic variations in the Earth's orbit—the Milankovitch cycles, which are explained in Chapter 9. These cycles have been an important driving force of climate change throughout geological time, giving rise to prolonged wet and dry spells tens of thousands to hundreds of thousands of years long. Under the influence of these climatic cycles, ice periodically melted rapidly over large areas far to the south. This added enormous volumes of water to the oceans and temporarily raised sea level by many tens of metres. The rising sea levels would have backed up river drainage, and bodies of brackish (partly salty) water submerged large parts of the Maritimes Basin, including the forested peatlands.

Late Carboniferous conglomerate and sandstone forming the famous "Flowerpot" sea stacks at Hopewell Cape, NB. Modern action of the powerful Fundy tides has undercut the rocks to form the unusual shapes.

Brackish water suited neither typical freshwater nor typical marine species, but a variety of "opportunistic" creatures took advantage of these difficult conditions. These included microscopic foraminifers, tiny ostracods and other, shrimp-like, crustaceans, bivalves and fish—even sharks. Fossils of these creatures are preserved in dark, shelly limestones and shales that sometimes overlie the coals. These shelly beds are called "clam coals" by coal miners, and good examples can be seen at Joggins, Mabou Mines and Sydney, NS, and at Minto, NB.

When the great southern ice sheets began to expand again, large amounts of water were once more "locked up" as ice, and sea level fell. Rivers flowed to the retreating sea across emerging coastal plains that were, again, being colonized by peat-forming vegetation. Hence, these cycles of peat formation, drowning, emergence and recolonization are the events recorded as cyclothems. In the ensuing millions of years, the peat was converted to coal and the overlying channel sand to sandstone. Thus, coal miners often find channel or valley fills of sandstone resting directly upon the coal in the roofs of mines, especially in the Sydney Coalfield, NS. Miners must be careful that the coal-cutting machinery does not hit the sandstones, generating sparks that can ignite any methane gas from the coal.

Life in the Coal Age Rainforest

The late Carboniferous fossils of the Maritimes have inspired visitors for over a century. For example, on 30 July 1842, Sir Charles Lyell wrote a letter to his sister from Truro, NS:

> *"We have just returned from an expedition of three days to the Strait which divides Nova Scotia from New Brunswick, whither I went to see a forest of fossil coal-trees—the most wonderful phenomenon perhaps that I have seen.... This subterranean forest exceeds in extent and quantity of timber all that have been discovered in Europe put together."*

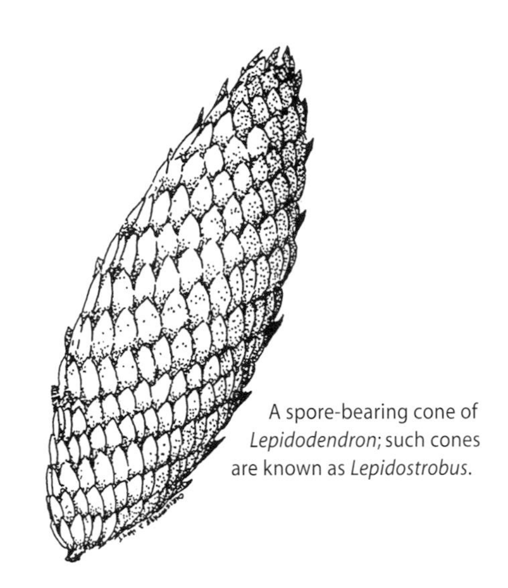

A spore-bearing cone of *Lepidodendron*; such cones are known as *Lepidostrobus*.

Even earlier, in 1836, Abraham Gesner had drawn attention to Joggins, NS, as "the place where the delicate herbage of a strange and now extinct vegetation is transmuted to stone."

Indeed, the Carboniferous plant life found at places like Joggins and the Cape Breton Coalfield seems strange to us, as most plants bore spores rather than pollen. There were no flowers or grass. Clubmoss trees, such as

A late Carboniferous *Lepidodendron* forest such as the one now preserved in the Sydney coalfield.

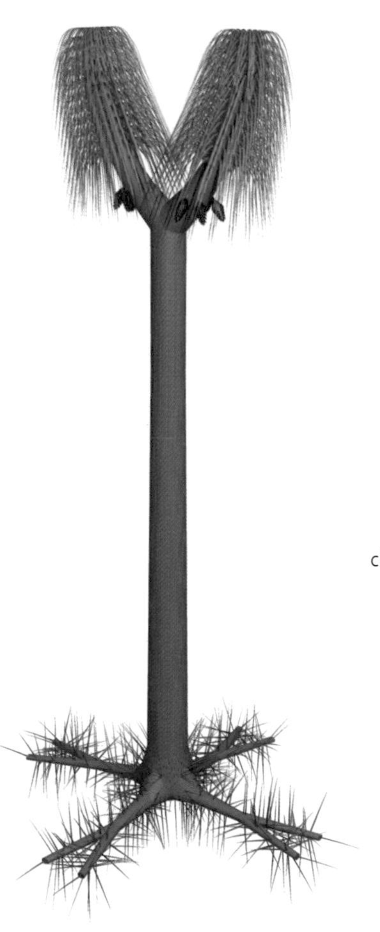

Reconstruction of a *Sigillaria* tree.

Fossil *Sigillaria* tree trunk in the cliff at Joggins, NS.

Lepidodendron and *Sigillaria*, with their long, dimpled roots called *Stigmaria*, towered unbranching to heights of 40 metres or more. The crowns of tree ferns arched gracefully above twisted, trunk-like roots. *Calamites*, with its jointed, bamboo-like stems and its narrow, whorled leaves, is one of the few Carboniferous trees with a closely related present-day survivor: the lowly scouring rush or horsetail, a miniature of its ancient forerunner. And there were seed ferns, a group of plants now extinct, which looked like true ferns but multiplied by producing seeds rather than spores. These were the trees of the Carboniferous coal swamps. Other plants clung to the ground or climbed, vine-like, on the trees. The ancestral conifer *Cordaites*, with

Whorl-shaped leaves known as *Asterophyllites*. This is the foliage of *Calamites*. Several branches with leaves are preserved on this specimen from late Carboniferous rocks near Clifton, NB.

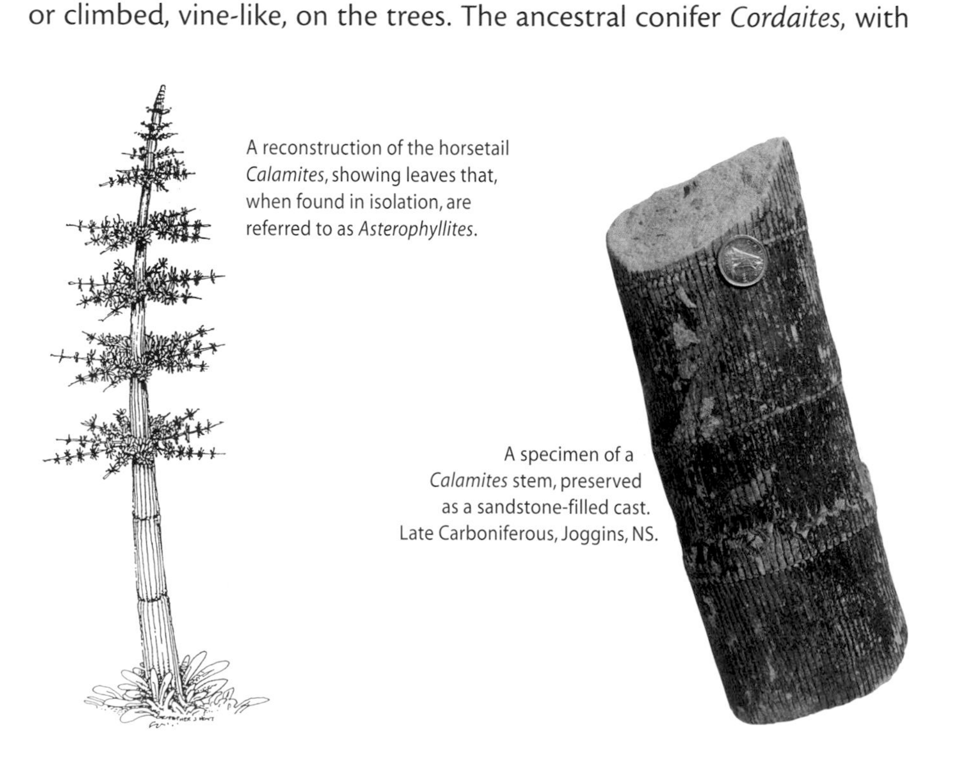

A reconstruction of the horsetail *Calamites*, showing leaves that, when found in isolation, are referred to as *Asterophyllites*.

A specimen of a *Calamites* stem, preserved as a sandstone-filled cast. Late Carboniferous, Joggins, NS.

its large, strap-shaped leaves, was unusual for the time because it bore both pollen and seeds.

The coal-swamp environment nurtured some of the earliest reptiles and land snails, as well as small to gigantic insects. There was also the giant millipede-like creature *Arthropleura*–at two metres one of the largest land arthropods that ever lived. It is known mainly from fossil trackways, since body fossils are rare. At Joggins, at Cape John near River John, and near Point Aconi in Nova Scotia, and at Tynemouth Creek near Gardner Creek in New Brunswick, fossil trackways wind across long-past floodplains and through ancient stands of *Calamites*. The trace fossils of *Arthropleura* show that it lived mainly on the forest floor amid the trunks of towering clubmosses. But its trackways have also been found in river channels, where water would have been readily available.

Amphibians and reptiles also inhabited Coal Age forests. Some of the earliest known reptiles were discovered by accident in Nova Scotia in 1852. Sir Charles Lyell was visiting the young William Dawson. On an excursion to Joggins, these two eminent geologists found bones of amphibians in a

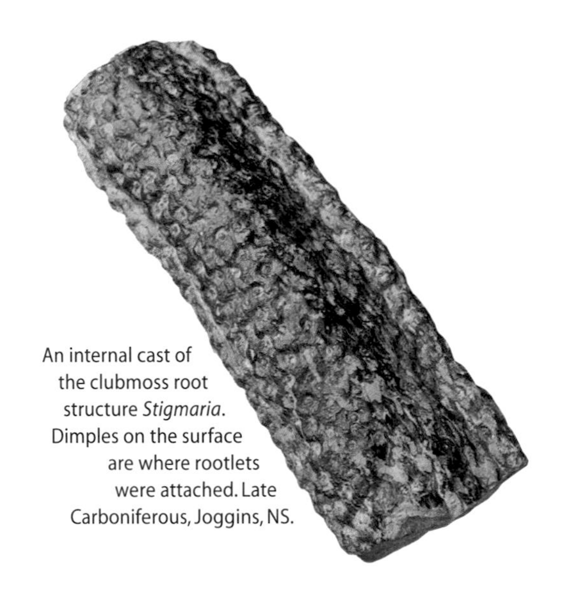

An internal cast of the clubmoss root structure *Stigmaria*. Dimples on the surface are where rootlets were attached. Late Carboniferous, Joggins, NS.

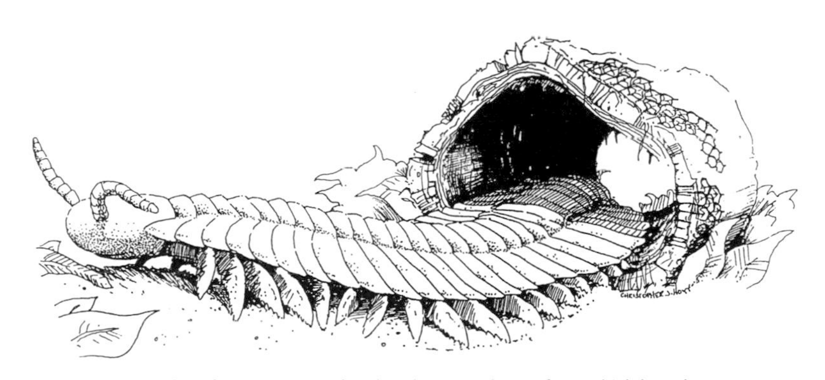

The Carboniferous myriapod *Arthropleura*, trackways from which have been found, for example, at Tynemouth Creek near Gardner Creek, NB and at Joggins, NS. Here, it is crawling out of the decayed and hollow centre of a clubmoss tree trunk.

Leaves and cones of late Carboniferous gymnosperm *Cordaites*.

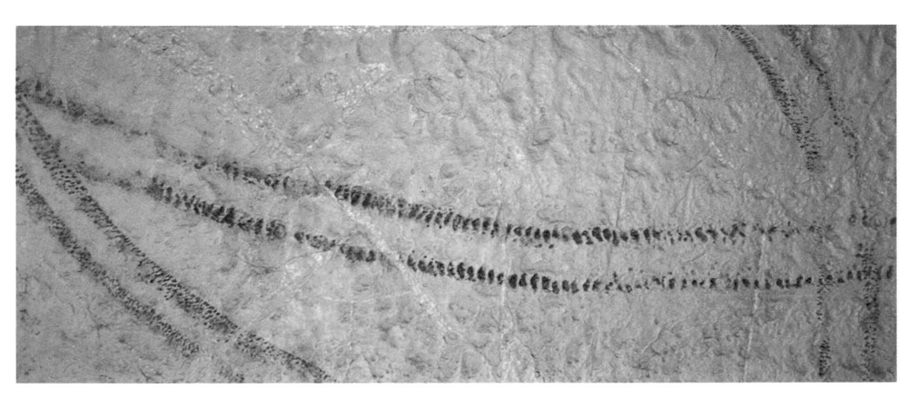

Artificial cast made from a series of trackways of the *Arthropleura* from late Carboniferous strata at Joggins, NS.

Jaws of the amphibian *Hylerpeton*, collected by Sir J.W. Dawson from late Carboniferous rocks at Joggins, NS.

segment of tree stump that had fallen from the cliff face. During later visits to Joggins, Dawson found the bones of over 100 individuals. Among these, he identified what was then the earliest known reptile: *Hylonomus lyelli*, or "Lyell's wood mouse", named by Dawson in honour of his earlier companion and mentor. Dawson and Lyell postulated that the animals had accidentally fallen into a hollowed-out tree stump when sediment accumulating around it had reached the stump's rim. Later, sediment from flooding rivers would have buried the animals within the tree trunk. Because the skeletons now lie mixed with charcoal within the tree stumps, a recent idea is that the animals may have entered the stump through "fire scars" and that they may have used the stumps as dens. However, it is also possible that the charcoal itself fell in over the rim of the stump. More than 150 years after their discovery, scientists are still asking questions about the fossils at Joggins.

The Great Drying, The Greatest Dying

The brick-red rocks of Prince Edward Island and the Northumberland Strait shores of New Brunswick and Nova Scotia are evidence of the "Great Drying" that heralded the end of the Coal Age. This event began about 295 million years ago, toward the end of the Carboniferous, and lasted through the Permian and into the Triassic. During this time, the Maritimes were still located within the supercontinent Pangea, which straddled the tropics. Hot air rose from this huge land mass that lay, baking, under the tropical sun. To replace this rising air (and depending on the season), moist air was pulled in from the vast surrounding Panthalassic ocean, or dry air was drawn from across the continent. The resulting climate of alternating wet and dry seasons has been described as a "megamonsoon".

Permian red beds exposed along the shoreline at Cavendish, PEI.

The Maritimes endured hot, arid conditions for several months each year, before the return of the monsoon rains. The persistent dry season spelled the end of the luxuriant wetland forests that had flourished in the Carboniferous. It also brought to a close the formation of peat (and thus coal) in the Maritimes. The peat-forming ecosystems of the Coal Age—centred on clubmoss trees with shallow rooting systems adapted to waterlogged soils—collapsed and eventually disappeared. The Permian dry-

Cliffs formed of Permian red beds at Lord Selkirk Provincial Park, PEI.

lands became home to trees tolerant of drought, notably the early conifers known as *Walchia*, which resemble the living Norfolk Island pine and have been found, for example, at Brule, NS. Plants from the Coal Age that grew in habitats outside the peatlands, such as *Calamites* and the tree-fern *Pecopteris*, still lived in wetter areas.

The Permian red beds of Prince Edward Island preserve a story of rivers swollen with muddy monsoonal floodwaters. These flowed in a northeasterly direction and shrank as the dry season set in. Buried in the pebbly river beds were scattered carcasses and bones of animals swept away in flash floods. In the 1850s, part of a skull and upper jaw of a reptile was

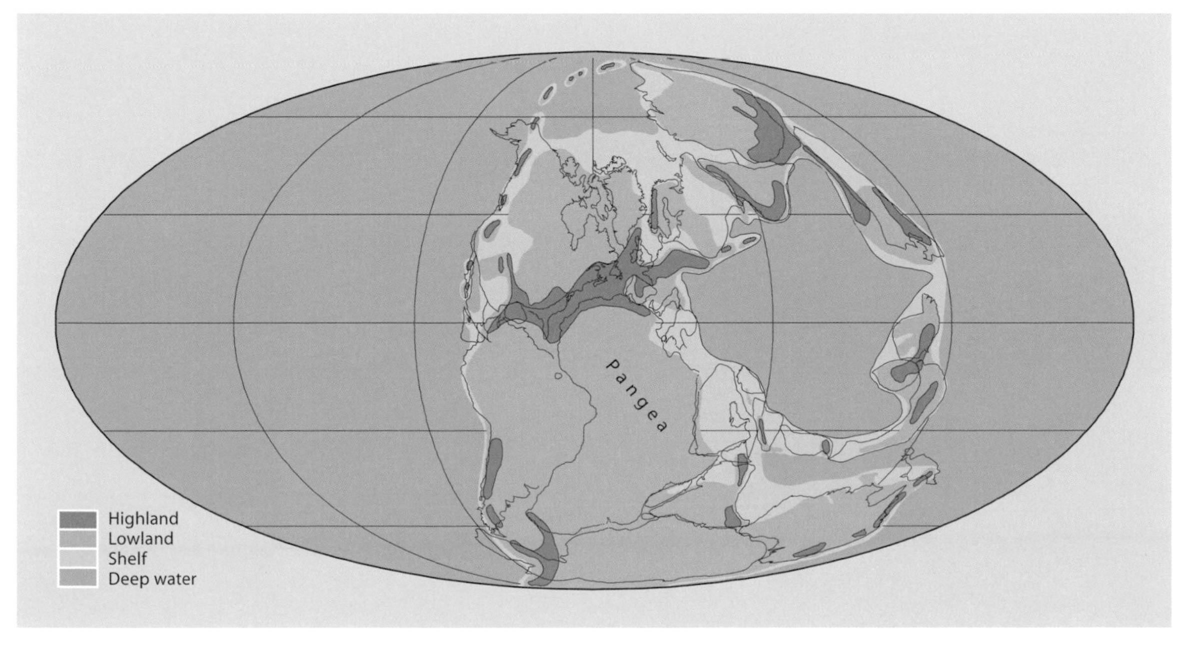

Global paleogeography of the Permian, about 255 million years ago.

A large skull fragment, with teeth attached, of *Bathygnathus*, found in Permian strata at New London, PEI, in the nineteenth century. This specimen, illustrated in Dawson's *Acadian Geology*, was once thought to be Canada's earliest-discovered dinosaur; but it is now known to be a mammal-like reptile, probably close kin to the sail-backed *Dimetrodon*.

discovered by a farmer digging a well near New London, PEI. The fossil was brought to the attention of William Dawson and sent to the American geologist Joseph Leidy, who named it *Bathygnathus*. Though the fossil is not complete enough to tell for certain, it may represent the earliest find of the fearsome 3-metre-long, sail-backed mammal-like reptile that is now known as *Dimetrodon*. (But even if it is not closely related to *Dimetrodon*, *Bathygnathus* is a mammal-like reptile, and not Canada's earliest-found "dinosaur", as was once believed.) This predator shared the landscape with the smaller, reptile-like *Seymouria*, its large plant-eating cousin *Diadectes*, and the nimble ancestors of modern lizards. The red beds of Prince Edward Island continue to yield exciting fossils. Trace fossils have been found at Point Prim. And, in 1997, a young boy discovered the complete skeleton of a reptile-like tetrapod while exploring the shore near his cottage at Miscouche.

By later Permian times, the parched Maritime landscape included Sahara-like sand dunes, piled up by the arid tropical winds of Pangea. The red rocks of the picturesque cliffs of the Magdalen Islands in the Gulf of St. Lawrence also bear witness to this much hotter and drier time.

No rocks from the latest Permian, about 250 million years ago, have been confirmed in the Maritimes. However, where rocks of this age are present in other regions, the fossil record bears witness to the greatest extinction in the history of life! About 90 percent of all species in the oceans perished. On land, about 60 percent of reptile and amphibian families and 30 percent of insect orders disappeared—the only mass extinction to have affected that group of six-legged invertebrates. The latest evidence suggests that this event took place within less than a million years. While not instantaneous in human terms, this time span is dramatic in a geological sense, given the enormity of changes that took place.

What caused this most dramatic of all extinction events? The later, more famous (but less devastating) extinction that claimed the dinosaurs is widely believed to have resulted in part from a cataclysmic meteorite impact. However, there is no evidence that an extraterrestrial impact triggered this earlier "Great Dying". As we have seen, Pangea's birth had created profound changes in climate, resulting in arid conditions and decline of vegetation, which must have caused a substantial drop in global oxygen levels. Toward the end of the Permian, there were apparently dramatic changes that we are only now beginning to understand. At first there was a major lowering of global sea level, then possibly a rapid rise. Added to this was a great increase in volcanic activity, especially in Siberia, which promoted a greenhouse effect by contributing carbon dioxide to the atmosphere. The result of all these sweeping changes over a short period of geological time was an extremely unstable climate and environment. Evidently, it was so inhospitable that only a small number of species managed to survive. Gone was the Paleozoic way of life. But this purge of the living world made way for an exciting and more active array of life-forms—among them the dinosaurs and early mammals—in the Mesozoic world.

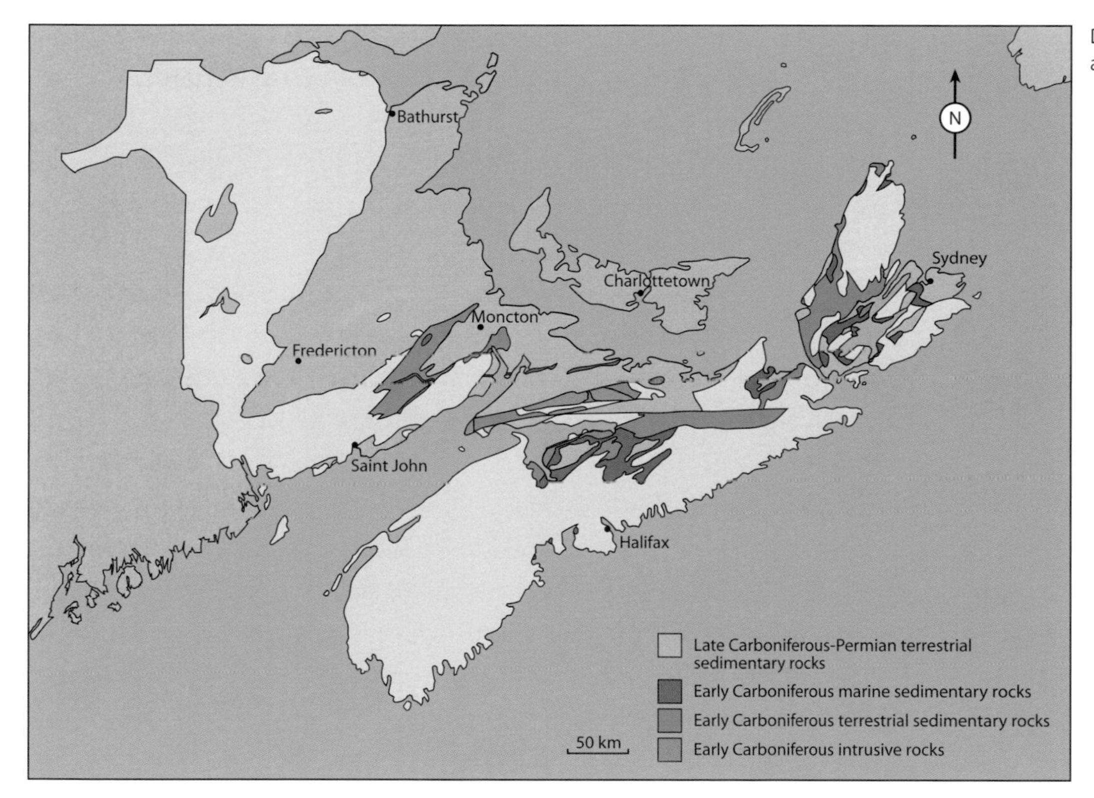

Distribution of Carboniferous and Permian rocks in the Maritimes.

Fish Tales: A History of Early Vertebrates

Well-preserved fish fossils are common in the Paleozoic and early Mesozoic rocks of the Maritimes. So it is fitting for a book on Maritime geology to have a section devoted to these fossils. As the earliest vertebrates, fish were first to evolve many critical features that we take for granted in our own bodies. For example, fish developed bone for the first time, as well as "tooth" enamel and dentine—perhaps we can blame fish for toothaches! Fish were the first vertebrates to have paired appendages, and their pectoral and pelvic fins eventually became arms and legs. Fish also developed jaws and, strange as it may seem, they evolved lungs, without which we could not breathe. In spite of our collective negligence in overfishing and in polluting their environment, fish are still the most successful vertebrate group on Earth, with almost 25,000 species alive today. How did this diversity come about?

One of the earliest known fish-like fossils is *Pikaia,* from the Cambrian Burgess Shale of British Columbia. The Burgess Shale is one of those rare deposits in which the soft parts of organisms have been preserved. The 5-centimetre-long *Pikaia* had a fish-like shape and a notochord (an internal rod-like support and the precursor of the vertebral column). The vertebral column is more popularly known as the "backbone", but this term is misleading since a vertebral column may be made of cartilage instead of bone. *Pikaia* also had repeating blocks of muscle known as myotomes. (When you check that a fish is properly cooked, the desired flakiness is caused by the separation of the myotomes.)

Although *Pikaia* was not a vertebrate because it lacked a true backbone, it did have a notochord and myotomes, so it was clearly related to vertebrates. Indeed, together with some other small animals that had notochords but no backbones, it can be grouped with vertebrates as a "chordate". All chordates have a spinal cord lying alongside the notochord, and this cord is enlarged at the front end of the animal to accommodate sense organs and at least a rudimentary brain. Within the chordates, the most successful group are the vertebrates, although even these represent less than five percent of all animal species living today. Fish were the earliest vertebrates, and their fossil record goes back to the Cambrian Period.

Yvonaspis, an armoured jawless fish that has been found in Devonian rocks at Campbellton, NB. Similar agnathan fish have been found in Devonian rocks near Arisaig, NS.

There is no single criterion that defines a fish, but paired fins, a vertebral column (in contrast to a notochord only) and bone are all important. Keys to a successful life in the water include speed and manoeuverability, both sideways and up and down. The development of fins was thus a major step in the evolution of fish. A pair of pectoral fins toward the front and a pair of pelvic fins toward the rear, together with the anal or tail fin, provide a fish with remarkable control over its speed and movement. Fins also have other major uses. The pectoral fins prevent sinking of the top-heavy front. And the pelvic fins, together with the anal and dorsal fins, stop the fish from rolling over in the water.

Yvonaspis, viewed from above.
A model from the New Brunswick Museum.

Bone is also vital. It provides rigid support for muscles and hence greater mobility. It is also a storehouse for some of the critical chemicals of life. And a bony external skeleton, so important to early fish, provides protection from predators such as nautiloids and the huge Paleozoic eurypterids.

The earliest fish are found in rocks about 530 to 470 million years old, from the Cambrian and Ordovician periods. These early fish, like modern lampreys and hagfish, were without jaws and had a more or less circular opening for a mouth. Thus, they are called "agnathans", a word derived from the Greek and meaning "without jaws". Agnathans lacked paired fins and had an internal skeleton of cartilage, instead of bone. However, starting in the Ordovician Period, many of them did have an outer covering of bony plates and scales, especially thick in the head region of the body. The flat shape of most of these early jawless fish suggests that they were bottom dwellers, feeding on organic debris. Jawless fish flourished and diversified during the Silurian and Devonian, but by the end of the Devonian most were extinct. Fish of these types are known from Silurian rocks at Nerepis, NB (*Cyathaspis* and *Thelodus*) and from Silurian and Devonian rocks near Arisaig, NS.

Small, tooth-like fossils called conodonts (meaning "cone-teeth") may represent agnathan fish. These microfossils, which range in age from late Cambrian to Triassic, are commonly used in biostratigraphy to date rocks in the Maritimes and elsewhere. Conodonts tend to occur in sets of elements, each set representing a particular conodont animal. For over a century, it was not known what kind of animal produced conodonts. Then, in the early 1980s, fossils of the entire soft body of conodont animals were found in early Carboniferous rocks in Scotland. They had worm-shaped bodies with myotomes, eyes, and tails with fins. These features, together with the observation that the conodonts themselves have a bone-like structure, suggest that the conodont animal was an early fish or a close relative.

The acanthodian fish *Climatius*, found in Devonian rocks at Campbellton, NB.

The Devonian placoderm *Phlyctaenius*, found in Devonian rocks at Campbellton, NB.

All fish have gills through which oxygen is absorbed from the water. Each pair of gills is supported by a pair of bones, called "gill arches". It appears that the first two pairs of gill arches evolved into jaws sometime before about 430 million years ago (during the Silurian Period). This was a major breakthrough in vertebrate evolution. Jaws—and the teeth that evolved with them—allowed access to a greater variety of food, provided predators with powerful weapons, and could also be used in defence. The development of jaws made it possible for fish to manage large food objects, permitting a change of diet away from filter feeding. To cope with changes wrought by the development of jaws and teeth, part of the gut became a tough sac, which we call a "stomach". During the Silurian and, especially, in the succeeding Devonian Period, there was a great diversification of jawed fish. So the Devonian, with its multiplicity of jawed as well as jawless fish, truly was the "Age of Fishes".

The first jawed fish were the acanthodians, or "spiny ones". Bone was still essentially limited to external scales and fin supports, but acanthodians were not heavily armoured, and this, together with a streamlined shape, suggests that they were active swimmers. Fossil acanthodians have been found in Devonian rocks at Campbellton, NB and in late Carboniferous rocks at both Joggins and Parrsboro, NS, and Chipman, NB. The placoderms were early jawed fish with heavy armour and, apparently, lungs. Some placoderms, the arthrodires, were up to 8 metres long and were voracious predators with sharp teeth. The arthrodire *Phylctaenius* has been found in Devonian rocks at Campbellton, NB. Another placoderm group, the antiarchs, had strange arm-like appendages. Fossils of some of these groups of fish have been found in the Devonian rocks of Miguasha, Quebec, just across Chaleur Bay from New Brunswick.

The Devonian antiarch *Bothriolepis*, found in Devonian rocks at Miguasha, Quebec, across the border from New Brunswick.

The vast majority of living fish, from cod and salmon to sea horses and sticklebacks, are the ray-

A palaeoniscoid ray-fin fish, similar to *Elonichthys*, from early Carboniferous rocks near Norton, NB.

An undescribed deep-bodied palaeoniscoid fish, similar to *Eurynotus*, found in early Carboniferous rocks at Albert Mines, NB.

finned fish—we will call them "ray-fins" for short. Ray-fins appeared in the late Silurian and are the earliest known true bony fish, with a bony vertebral column. They had a regular pattern of fins, all of which were supported by long, thin, flexible internal rods, or fin rays. This group of fish had pairs of pectoral and pelvic fins, supported by pectoral and pelvic girdles. Fossils of early ray-fins, known as palaeoniscoids, are found in various Carboniferous strata, including the Albert oil shales in New Brunswick. More advanced ray-fins, known as neopterygians, are represented by *Semionotus*, a thick-scaled freshwater dweller that lived in lakes of the Triassic and Jurassic Fundy Basin. The teleosts were the latest major group of ray-fins to evolve, and include most of today's familiar fish. In the Maritimes, however, teleost fossils are known only from the Quaternary.

Another evolutionarily important (but now rare) group of bony fish are the lobe-finned fish, so called because of their fleshy pectoral and pelvic fins. In one group of "lobe-fins", the crossopterygians, these fins had an internal skeleton rather

Fossil of a freshwater ray-fin fish, *Semionotus*, of early Jurassic age. Five Islands, NS.

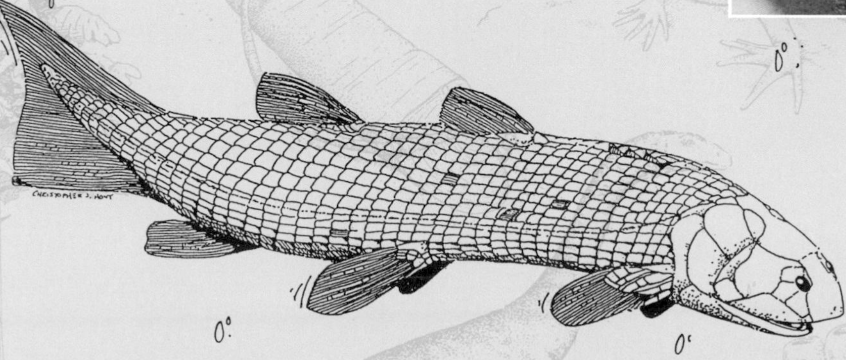

The Devonian crossopterygian fish *Latvius*, found in early Carboniferous rocks north of Moncton, NB.

The early Devonian shark *Doliodus*, remains of which have been found at Campbellton, NB.

than fin rays for support. This skeleton was the foundation for amphibian limbs, and eventually, therefore, for those of reptiles and mammals. Fossil crossopterygians have been found in early Carboniferous strata at Blue Beach near Hantsport, NS and near Moncton, NB, including the genus *Latvius*. Crossopterygians were thought to be extinct until a living form, the coelacanth, was caught in deep water off Madagascar in 1938. Lungfish are another group of lobe-fins, so named because their modern representatives use lungs as well as gills. The variety of modern lobe-fins pales beside the diversity of fossil forms, some of which had skulls so similar to those of amphibians that they were described as such until the rest of their skeleton was known. Some were predators up to several metres long. Their huge fang-like teeth, up to 4 centimetres in length, are occasionally found in Carboniferous rocks in the Maritimes.

The final major group of fish is the cartilaginous fish: sharks, rays and skates. Like early fish, their internal skeletons are made mostly of cartilage. Sharks have no external bony armour, either, so their fossils are usually restricted to teeth and fin spines. Sharks likely evolved from jawless fish known as thelodonts, examples of which have been found in Silurian rocks near Nerepis, NB. A specimen of the early Devonian shark *Doliodus* from Campbellton, NB, is the oldest known complete fossil shark. Teeth of freshwater sharks are found at Joggins and Pictou, NS; at Chipman, NB; and at other Carboniferous sites. Large teeth of more modern sharks have been dredged up by fishers' nets on Georges Bank.

"Jaws Plus": the tooth of a large extinct shark from possible Miocene strata. Georges Bank.

CHAPTER 7
An Ocean is Born

A scene in the Fundy Basin in late Triassic time, about 210 million years ago.
The small dinosaurs are carnivores, probably of the genus *Syntarsus*.

Break-up of Pangea

North America really is a chip off the old block of the supercontinent Pangea, which began to break up about 250 million years ago. This break-up has formed the modern continents and oceans, with the exception of the Pacific Ocean. First, Pangea started to divide into a northern part, Laurasia, and a southern part, Gondwana; these were partially separated by an ocean known as Tethys, named after the wife of the Greek god Oceanus. However, before Tethys completely rent Laurasia from Gondwana, the nascent North Atlantic Ocean began to separate what was to become North America from Eurasia (Europe and Asia). Lastly, within Gondwana, southern Africa and South America separated, forming the South Atlantic Ocean. But before we tell the full story of the birth and evolution of the Atlantic Ocean, we need to examine Triassic and Jurassic rocks—those about 230 to 200 million years old—in and around the Bay of Fundy.

The Fundy Basin

During the Triassic and early Jurassic, about 250 to 190 million years ago, the area of the present-day Maritime Provinces was still within Pangea. As in the Permian, the Maritime region in the Triassic was mountainous and farther south than at present, with a warm tropical climate. During the late Triassic, Pangea drifted northward and, in the latest Triassic and early Jurassic, the Maritime climate changed from warm and humid to warm and arid. This affected vegetation, which started out relatively luxuriant but, by the latest Triassic, became sparse and concentrated on the shores of rivers and shallow intermittent lakes.

Beginning in the early Triassic, the northern part of Pangea was rent by fissures that grew into a rift valley system, similar to today's East African Rift Valley and Red Sea. As rifting thinned and stretched the crust, a series of sedimentary basins formed, especially near lines of weakness in the crust. One such line of weakness was the Cobequid-Chedabucto Fault System, and it was in the vicinity of this fracture zone that the Fundy Basin developed, located where the Bay of Fundy is today. The sediments that filled the ancient Fundy Basin came mainly from the west, where wind and water were eroding the already-ancient Appalachian Mountain chain.

Fundy Basin strata can be seen, for example, in the red cliffs at Blomidon and Five Islands, NS. Some of the sediments (now sedimentary rocks) were river deposits, whereas others represent wind-blown sand dunes, alluvial fans at the edge of the basin, and lake beds. Indeed, the central

Global paleogeography of the late Triassic, about 215 million years ago.

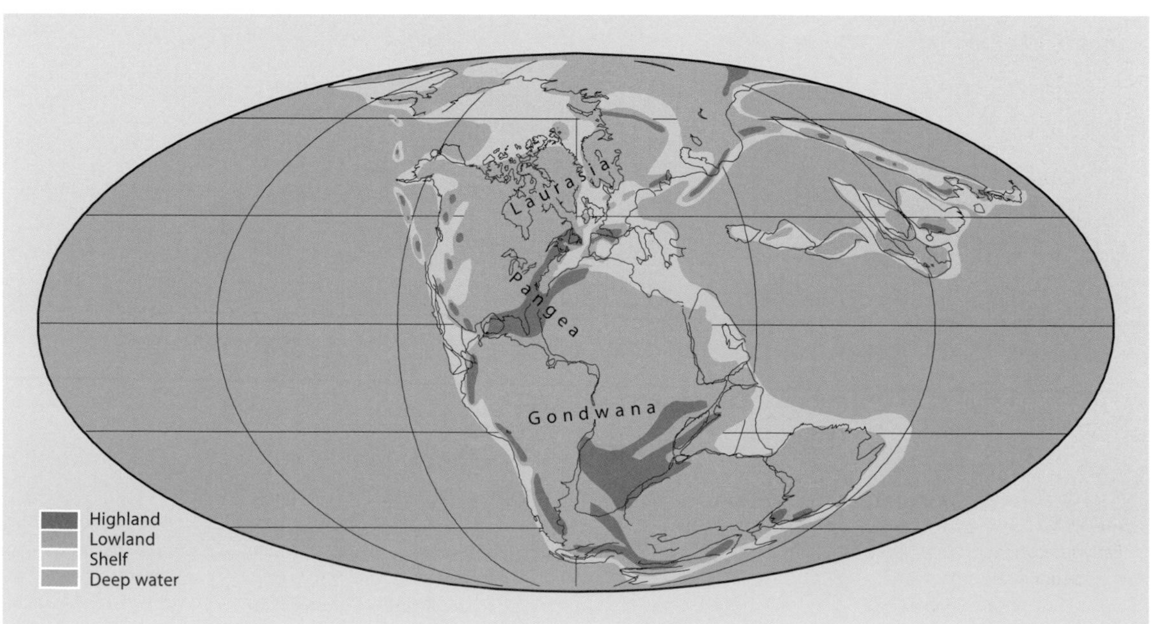

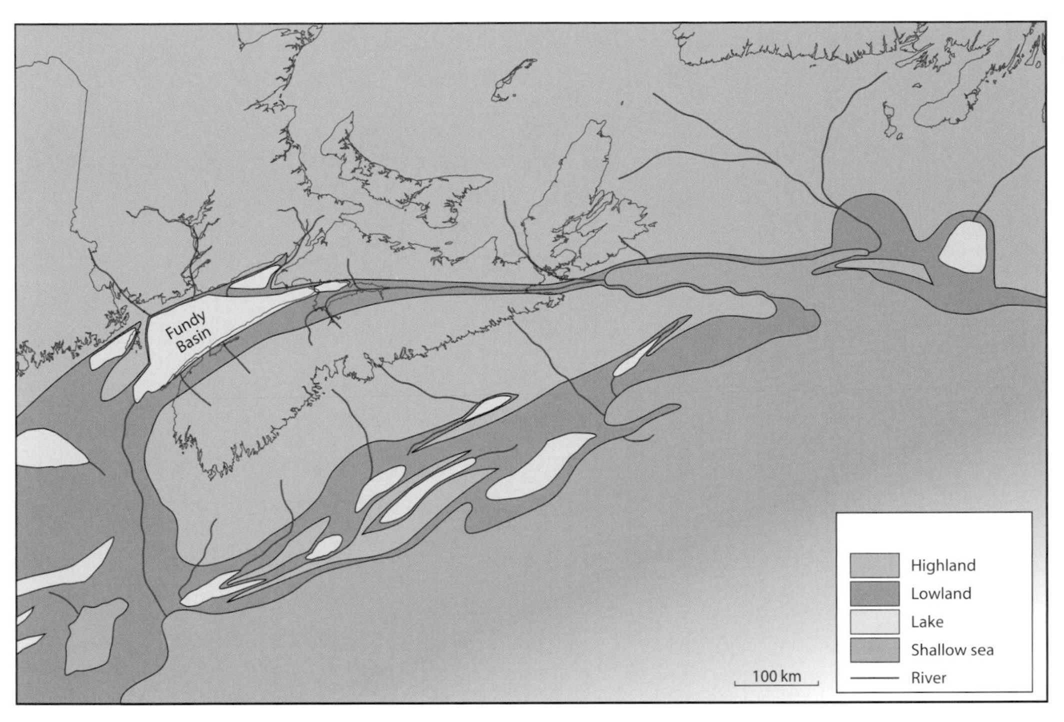

Regional paleogeography about 200 million years ago.

part of the basin was often occupied by one or more large salt lakes. These dried up from time to time, leaving behind salt and gypsum deposits within the thinly bedded, flat clastic layers so typical of lake deposits. The presence of salt and gypsum confirms that the climate was hot and dry. Oil and gas exploration wells drilled in other Triassic rift basins on the Scotian

Early Jurassic sand dunes preserved as cross-bedded sandstone at Wasson Bluff, near Parrsboro, NS.

Late Triassic sandstone at Red Head, near Lower Economy, NS. The cross-bedded sandstone in the lower part of the cliff is a relict of wind-blown sand dunes. The flat-bedded sandstone in the upper part of the cliff represents deposits of ephemeral lakes.

This grey cliff at St. Martins, NB, is made up of a Triassic conglomerate deposited as an alluvial fan at the edge of the Fundy Basin. The red rocks in the foreground were formed in a river system and are probably of Permian age.

Shelf (discussed below) contain similar rocks and therefore indicate similar conditions.

During continental break-up, magma often moves to the surface through deep fissures in the crust and is erupted as lava flows, as pyroclastics (volcanic ash and bombs) and as gases. Such igneous activity took place in the Fundy Basin in earliest Jurassic time. Indeed, these Fundy Basin rocks are part of the largest known episode of volcanic activity on Earth. Identical rocks of the same age are found along the edges of post-Pangean continents from Brazil to France and from Fundy to Morocco.

Basaltic lavas from this igneous activity now form landscape features in and around the Bay of Fundy, including the North Mountain of Nova Scotia, which extends from Cape Blomidon near Scots Bay to Brier Island. This North Mountain Basalt underlies most of the Bay of Fundy and covers much of Grand Manan Island, NB. It is also present along the north shore of the Minas Basin in Nova Scotia, at Cape d'Or near Advocate Harbour in the west, around Parrsboro, and at Economy Mountain near Five Islands in the east. A common feature of basalts is columnar jointing, which produces tall polygonal columns that form as lava cools, solidifies and shrinks. Columnar basalts can be seen on Brier Island, NS; on Grand Manan Island, NB; and elsewhere around the Bay.

Early Jurassic columnar basalt on Brier Island, NS.

Latest Triassic lake sediments at Blomidon, NS.

The tops of many lava flows are "frothy"—riddled with holes caused by escaping gases as the lava cooled. The holes have since been partially filled with minerals precipitated by seeping ground water. These minerals include not only common types like calcite and rock crystal quartz, but rarer minerals much sought by "rock hounds". Among these rarer minerals are amethyst, agate, jasper and the zeolites heulandite, chabazite and stilbite, the last being Nova Scotia's provincial mineral.

Erosion near Scots Bay, NS, has revealed a section through columnar basalt.

Rift-related basalts are also found as long, linear features (dykes) formed within the crust. These dykes were originally fissures through which magma rose to the surface to form lava flows. Examples are the Shelburne Dyke along the South Shore of Nova Scotia, the Caraquet Dyke that extends from northeastern New Brunswick to eastern Maine, Minister Island Dyke on Minister Island (near St. Andrews, NB), and an unnamed dyke exposed at Malpeque Bay, north of Summerside, PEI.

There are no known sediments younger than early Jurassic in the Fundy Basin, presumably because rifting stopped and consequently the area no longer formed a depression in the Earth's crust in which sediments could collect.

A large pebble of early Jurassic basalt showing its vesicular nature; the vesicles are partially infilled, mainly by heulandite. Halls Harbour, NS

Fossils of the Fundy Triassic

To the casual observer, the Triassic and Jurassic rocks of the Bay of Fundy shoreline seem almost devoid of fossils and, until a few short decades ago, paleontologists would have agreed. However, thanks to the efforts of a few diligent searchers—among them paleontologist Paul Olsen and fossil hunter Eldon George of Parrsboro, NS—many fossil treasures have now been discovered along the Bay's shoreline. These treasures include Canada's oldest dinosaurs by tens of millions of years at Burntcoat Head, the world's smallest dinosaur footprints and, at Wasson Bluff near Parrsboro, one of the richest troves of earliest Jurassic vertebrate fossils ever found. The age of the fossils at Wasson Bluff make them particularly special, as we will explain shortly.

Before turning specifically to the fossil finds of the Fundy shores, it is helpful to review the evolution of reptiles. As discussed in Chapter 3, reptiles arose from amphibians in the Carboniferous, evolving eggs that could be laid away from water. As we saw in Chapter 6, some of the earliest

Part of a "rubbly" basaltic lava flow at Wasson Bluff, near Parrsboro, NS. The cavities in the basalt have been partially infilled by orange chabazite crystals.

Tanystropheus, a Triassic aquatic reptile.

reptiles were discovered in Carboniferous tree trunks at Joggins, NS, which is also on the shores of the Bay of Fundy. From details of their skeletons, experts believe that early reptiles diverged into two lineages. The first, the mammal-like reptile lineage, led eventually to mammals—including us. The second led to all other reptile groups, including lizards, snakes, turtles, crocodiles and dinosaurs. We refer to the creatures of this latter lineage as mainstream reptiles.

The mammal-like reptiles first appeared in the late Carboniferous, just over 300 million years ago, and became so successful that they were the dominant land vertebrates in Permian to middle Triassic times. One example was the possibly sail-backed *Bathygnathus* from the Permian of Prince Edward Island, and mammal-like reptiles probably made many of the footprints in the Permian *Walchia* forest at Brule, NS.

Along with most other forms of life, mammal-like reptiles were decimated in the great Permian dying. Early Triassic rocks and fossils are not present in the Maritimes, but from evidence elsewhere, we know that the earliest Triassic faunas contained few species, and that these few had widespread distributions; the few survivors of the dying were evidently having their "day in the sun". Mammal-like reptiles were important among these survivors.

By the middle Triassic, recovery from the end-Permian extinction event was proceeding apace. Among the mammal-like reptiles, there were new forms of dicynodont (animals with "two dog teeth"); these were heavy herbivorous animals with horny beaks, as well as two large canine teeth, which may have been used in display. Fossils of dicynodonts have been found in middle Triassic rocks at Carrs Brook, NS. Another group, the cynodonts (animals with "dog teeth") are represented by *Arctotraversodon* from Burntcoat, NS. This was a bizarre mammal-like reptile that looked like a massive, bear-sized, sabre-toothed pig.

There are also fossils of mainstream reptiles in the middle and late Triassic beds of the Fundy Basin. These include *Hypsognathus*, a 30-centimetre-long, buck-toothed reptile from Paddys Island, near Medford, NS; *Tanystropheus*, an aquatic reptile, which was up to 7 metres long—including a neck twice as long as its body—from Carrs Brook, NS; and pig-like rhyncosaurs, specimens of which have been found at Evangeline Beach, near Grand Pré, and at Burntcoat, NS. There were also a selection of crocodile-like reptiles, one group of which, the rauisuchids, have left remains of their 6-centimetre-long teeth at Carrs Brook and Burntcoat.

Both dinosaurs and mammals first appeared in the later Triassic. There is no evidence for mammals in the Fundy Basin, but there are dinosaur bones and footprints. For example, three-toed, bird-like tracks (known as *Anchsauripus*) from Rossway, NS, may have been made by the predatory *Coelophysis*. This bipedal dinosaur was up to 2 metres long, apparently lived in herds and had many razor-sharp teeth. Tracks known as *Atreipus* have been found at Paddys Island. (Because the maker of a fossil trackway is rarely identifiable, at least with confidence, each type of trackway has been given its own Latin name. Thus, the name *Atreipus* refers to the fossil trackway, not to the animal that made it.) *Atreipus* tracks are dog-like prints, with tiny claws and four prominent toes, and were probably made by small, bird-hipped dinosaurs.

Early Jurassic rocks at Wasson Bluff, near Parrsboro, NS, where reptile bones (including those of dinosaurs), shark teeth and other fish remains have been found. The bulge of dark grey rock in the right half of the photo is a basaltic lava flow. The small, white area represents lake sediment, and the red rock to the left represents sedimentary rocks formed in a river. The light grey rubbly area in the centre represents ancient talus.

Jurassic Bluff

Evidence from fossil pollen and spores and from radiometric dating has shown that the Triassic-Jurassic boundary lies just below the base of the North Mountain Basalt. This boundary marks another major extinction event, in which about 40 percent of land vertebrate families were wiped out. This is less dramatic than the Permian dying, but more devastating than the famous Cretaceous-Tertiary event associated with the demise of the dinosaurs. Indeed, it was this end-Triassic extinction event, about 200 million years ago, that seems to have provided the opportunity for dinosaurs to expand and dominate life on land for the next 140 million years. Gone were the crocodile-like reptiles (although, of course, true crocodiles survived) along with most of the mammal-like reptiles. The "Age of Dinosaurs" had arrived.

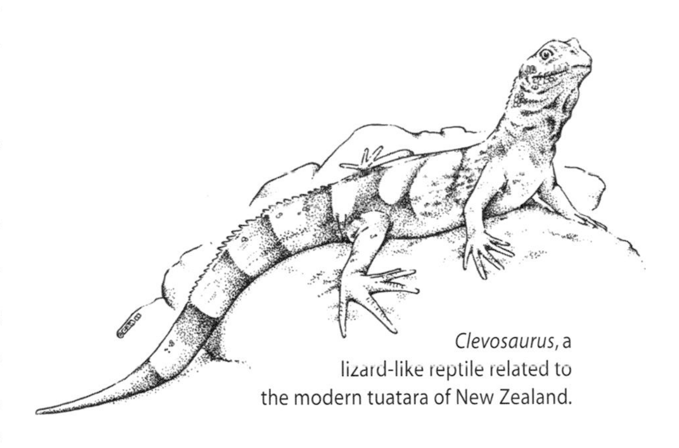

Clevosaurus, a lizard-like reptile related to the modern tuatara of New Zealand.

Therefore, the lava flows of the North Mountain Basalt and the sedimentary rocks associated with them are of earliest Jurassic age, a highly significant time in the history of life. This explains the excitement when, in early 1986, a find of "100,000 fossils" from Wasson Bluff, near Parrsboro, was reported by the international news media). In the summer of 1984, inspired by earlier finds, paleontologists Paul Olsen, Neil Shubin and Hans Dieter Sues were searching the cliffs at Wasson Bluff, when Shubin's eyes picked out hundreds of small, white bone fragments.

Side view of the skull of *Clevosaurus*, from Wasson Bluff, near Parrsboro, NS.

The assemblage at Wasson Bluff contains bones from several different types of animal. These include the lizard-like reptile *Clevosaurus*, whose closest living relative is the lizard-like tuatara

Protosuchus, a long-legged, early Jurassic crocodilian.

Lower jaw of the "sabre-toothed" crocodile, *Protosuchus*, from Wasson Bluff near Parrsboro, NS.

of New Zealand. There was the "sabre-toothed" *Protosuchus*, sometimes called the "cheetah of the Jurassic", a long-legged, short-snouted crocodilian that was up to about 50 centimetres in length and that had sharp teeth, two of which were greatly enlarged. It perhaps preyed upon *Clevosaurus* and small dinosaurs. There were also specimens of mouse-sized *Pachygenelus*, a mammal-like reptile so closely similar to mammals that it may have been furry and warm-blooded.

The rocks at Wasson Bluff represent a variety of early Jurassic environments, adding to the site's fascination. There are the basaltic lava flows, as well as preserved mud and boulder flows representing volcanic avalanches (or "lahars"). There are sedimentary rocks that represent river and lake deposits, as well as wind-blown sand dunes and scree deposits (or "paleotalus"). Many of the fossils are remains of animals that fell into fissures, especially in the paleotalus deposits. There are also fossil fish in lake sediments at Wasson Bluff, as well as at Five Islands and, across the Minas Basin, at Scots Bay.

There may be no mammals in the early Jurassic of the Fundy Basin, but there are dinosaurs. For example, several incomplete skeletons of prosauropods at Wasson Bluff have been found in rocks that originated as wind-blown dune sands. Prosauropods (animals "before the lizard feet") were usually only 2-5 metres long, but gave rise in later Jurassic times to the sauropods (animals with "lizard feet"), the largest land animals that have ever existed. Sauropods included such leviathans as *Diplodocus*, *Brachiosaurus* and the "ground-shaker" *Seismosaurus*. Unlike four-footed sauropods, prosauropods—such as those found at Wasson Bluff—probably walked mainly on two legs. However, like sauropods, they were probably plant-eaters. Prosauropod footprints have also been found in the Fundy Basin.

Dinosaur fragments at Wasson Bluff and elsewhere in the Fundy Basin show that there were other types of dinosaurs here. Some teeth and bones may have belonged to *Syntarsus*, an agile, carnivorous dinosaur, up to 3 metres long. This creature had a small head with numerous serrated teeth (good perhaps for cutting meat), a long neck, a sleek body, a whip-like tail, and arms and legs with claws ideal for slashing and stabbing. Bones and teeth of a *Lesothosaurus*-like dinosaur have also been found.

Pachygenelus, an early Jurassic mammal-like reptile.

Many dinosaurs are considered to have been cold-blooded, like today's reptiles and amphibians. But active, carnivorous types like *Syntarsus* may have been warm-blooded. An elevated, stable body temperature would have been a tremendous advantage to a carnivore, enabling it to outrun its prey and to be active at any time of day or night. It would, however, have come at the cost of greater food requirements. *Syntarsus* ran on its two hind legs, rather than on all fours, which also contributed to its effectiveness as a predator. This bipedal mode was common in carnivorous dinosaurs, even in the gigantic *Tyrannosaurus*, which roamed the Earth more than 100 million years later.

The vegetation of the Triassic and Jurassic world would have been more familiar to modern eyes than that of the Carboniferous coal swamps. But there were still no flowering plants. The dominant plants were gymnosperms (especially conifers, but also palm-like cycadeoids, cycads, and ginkgos—a modern example of which can be found in Halifax's Public Gardens). In addition to gymnosperms, there were horsetails, seed ferns and true ferns, some of which developed into trees. Conifers, the largest plants, included forms related to today's giant sequoias, pines and monkey puzzle trees. Occasional remains of conifers have been found in early Jurassic lake sediments—for example, at Ross Creek near Scots Bay, NS, where these remains are heavily encrusted with stromatolite layers.

Perhaps the most surprising difference from the modern landscape was the lack of grass, which did not evolve for another 150 million years. This meant that herbivores existed on a diet of ferns, shrubs and small trees, very different from what today's grazers eat.

The early Jurassic prosauropod dinosaur *Anchisaurus*.

The Nascant Atlantic

The story of the early Atlantic Ocean is recorded in sediments deposited offshore from Nova Scotia. During the Triassic and early Jurassic, as rifting thinned and stretched the crust to form the Fundy Basin, the same processes produced a series of sedimentary basins to the south and east of Nova Scotia, where the Scotian Shelf and Slope are now located. The sediments that filled these Scotian margin basins, like those in the Fundy Basin, came mainly from erosion of the Appalachian Mountains. The oldest sediments in these nascent Atlantic rift basins were sands and muds deposited in seasonal rivers and intermittent lakes, and are commonly preserved as red beds, so characteristic of nonmarine sedimentary rocks. Some areas were then flooded from east to west by a shallow arm of the Tethys Ocean. The hot climate periodically evaporated the sea water, leaving thick deposits of salt.

The Atlantic Ocean slowly opened, further fragmenting Pangea and, in the process, splitting in two the area that had once been the microcontinent of Armorica. Eastern Canada drifted away from what is now southern Europe and northwest Africa, with Nova Scotia retaining the part of offshore Armorica now known as the Meguma Terrane. As this happened, the basins southeast of Nova Scotia expanded and joined to form one large, marine basin called the Scotian Basin, underlain by rocks of the Meguma Terrane. The Scotian Basin now contains sedimentary rocks up to 20 kilometres thick. Throughout most of the early and late Jurassic, the majority of sediments being deposited in the basin were clas-

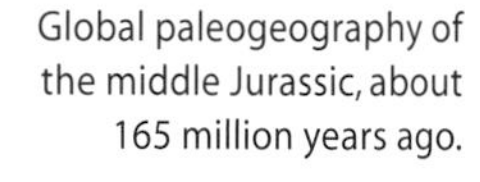

Global paleogeography of the middle Jurassic, about 165 million years ago.

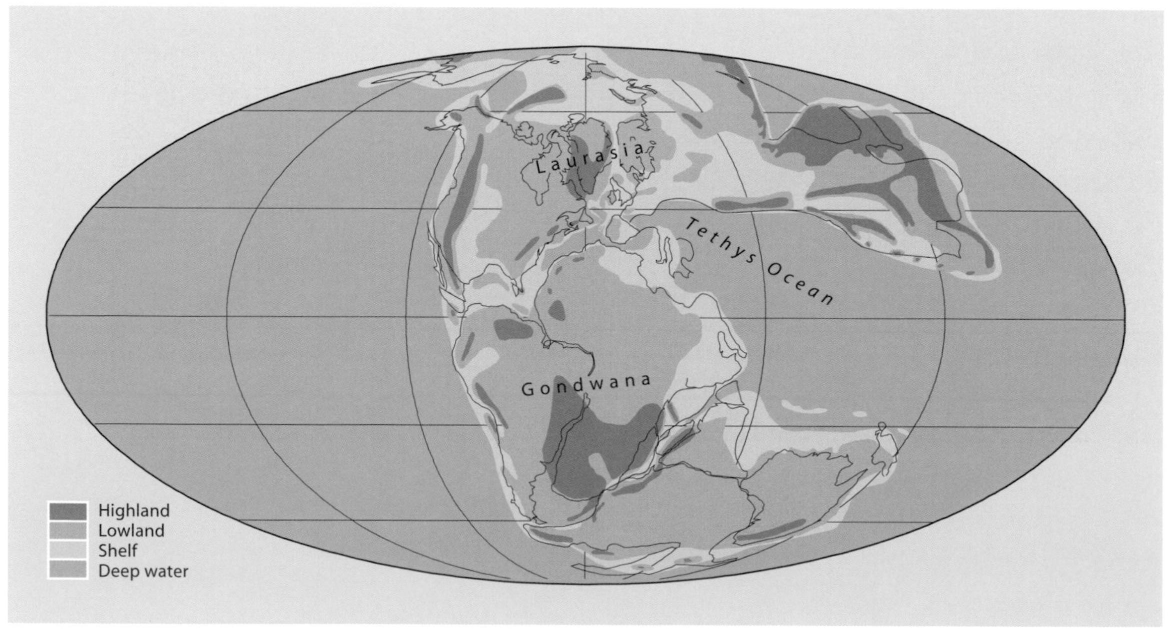

Sable Delta
Laurentian Delta
Shelburne Delta
100 km
Highland
Lowland
Delta
Shallow water
Limestone bank
Deep water
River

Regional paleogeography in the late Jurassic, about 150 million years ago.

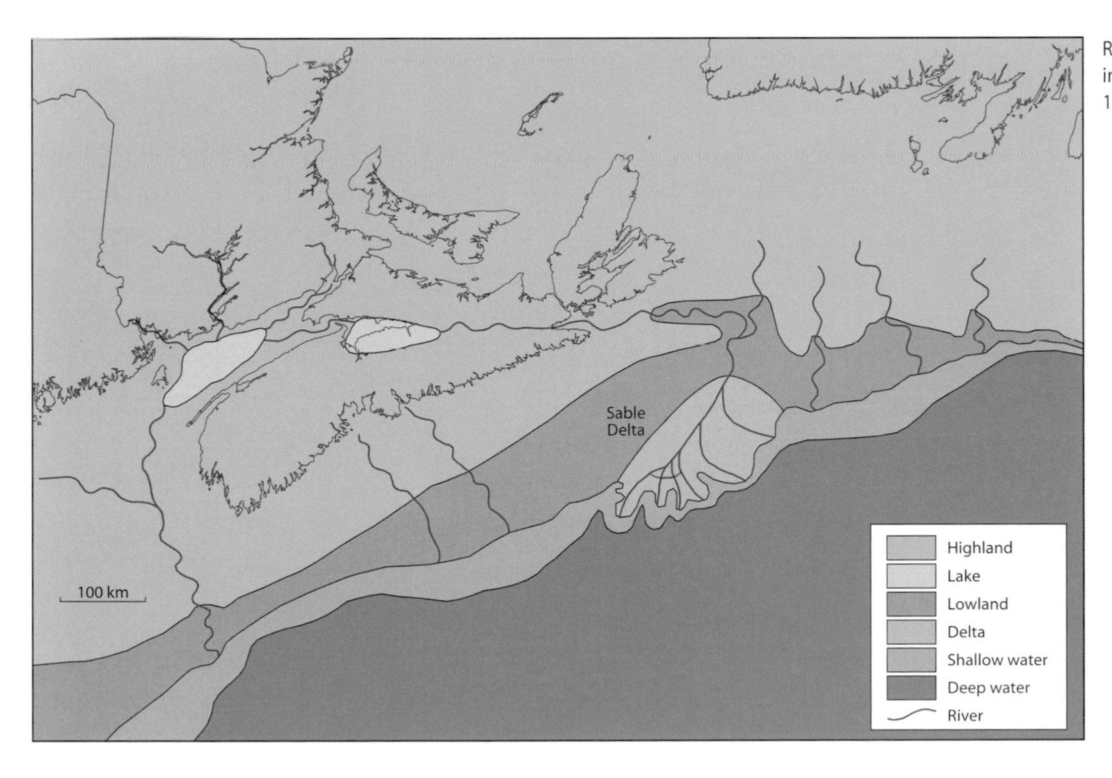

Regional paleogeography in the early Cretaceous, about 135 million years ago.

tics, sands closer to shore and shales farther offshore. About 150 million years ago, limestone was deposited in parts of the Scotian Basin, further supporting the idea that the Maritimes had a warm climate during the Jurassic. Reefs flourished in the shallow seas off eastern Nova Scotia, just as they do today in the Bahamas.

Periodically, when sea level fell, land extended 100 kilometres or more to the south of present-day Nova Scotia. At such times, parts of the area now occupied by the Scotian Shelf were vegetated swampy lowlands, probably inhabited by dinosaurs and other reptiles. But during times of high sea level, waves eroded sea cliffs in the Meguma rocks off southern Nova Scotia, only a few kilometres from today's coastline.

The area of greatest sediment deposition in the Scotian Basin in the late Jurassic was at the mouth of the ancestral St. Lawrence River, the site of two major deltas, Laurentian and Sable. Rivers draining Quebec and the Maritimes contributed sand and mud that accumulated to a thickness of several thousand metres in the Scotian Basin.

Age of the Chalk Seas

The Cretaceous Period, from about 140 to 65 million years ago, was a time of generally warm climates worldwide. As the area of present-day Newfoundland and Labrador began to separate from Europe, the Atlantic Ocean extended northward. Eastern North America was beginning to take on its modern appearance. The ancestral St. Lawrence River, now

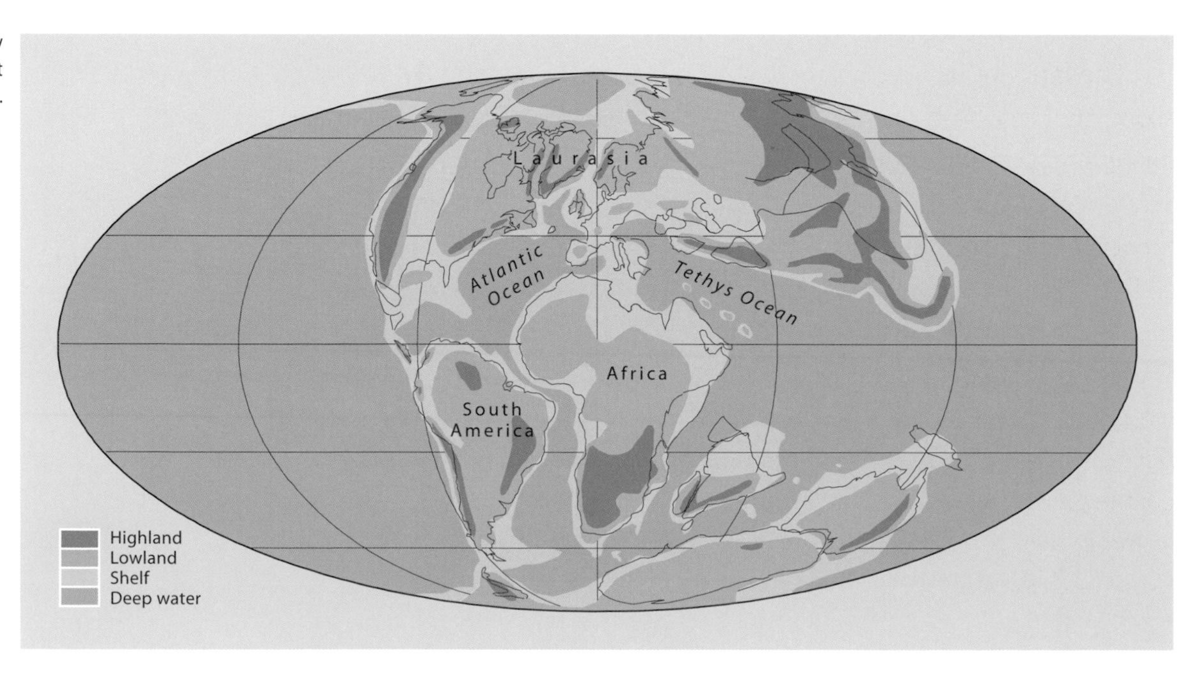

Global paleogeography of the Cretaceous, about 100 million years ago.

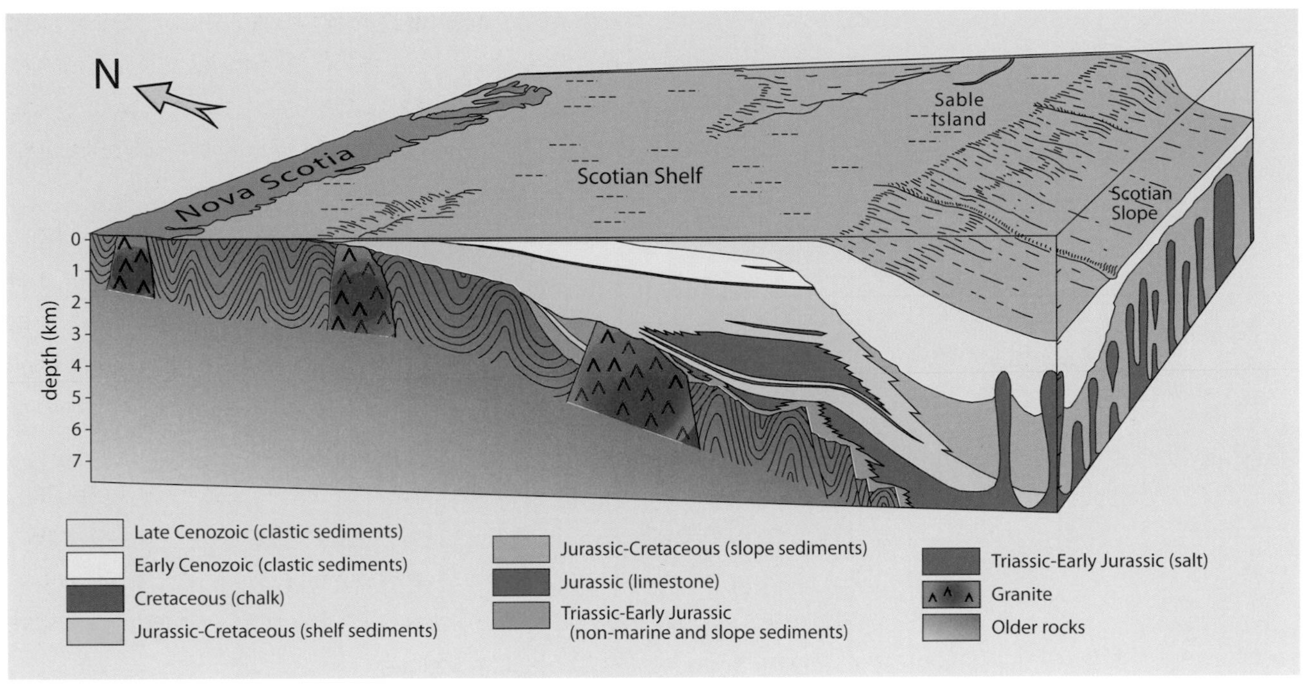

The general geology of the Scotian Basin.

appears to have had a single delta, the Sable Delta, centred in the vicinity of Sable Island. It was still supplied, in part, with sediments from the ongoing erosion of the Appalachians. But conditions affecting the basin began to change. As mountainous areas were worn down, rivers became more sluggish, and finer sediment—mainly silt and mud—was transported into the Scotian Basin. Meanwhile, as sea level rose, the shoreline was slowly migrating northwestward toward modern Nova Scotia, and marine sediments were once again deposited across the shelf.

The late Cretaceous was a time of extensive limestone deposition, especially of a type known as chalk. The white cliffs of Dover in England are formed from chalk, and this type of sediment was also deposited on the Scotian Shelf. Chalk is composed almost entirely of microscopic shell fragments called "coccoliths". Since the microfossils producing coccoliths generally live in warmer waters, the widespread distribution of chalk shows that much of the Earth's climate was tropical to subtropical during the late Cretaceous. Besides coccoliths, the Scotian Shelf sedimentary rocks yield many other types of microfossils, such as foraminifera and dinoflagellates. Because of the profusion and rapid evolution of these microfossils, they enable geologists to date rocks from offshore wells. Larger fossils also are found in well cores and consist mostly of clams and other shellfish, trace fossils and plant impressions.

A "pit", or quarry, near Shubenacadie, NS, exposing white Cretaceous river sands that are extracted for use in glass-making, sand blasting, water treatment and other applications. A brown clay till (a deposit of the last ice advance over the Maritimes) lies above the sand, and in places has collapsed over it. The exposure above the upper roadway is a waste dump, mostly of clay till. The sand, although about 100 million years old, is loose—it has not yet been consolidated into a sandstone.

Back on Land

Onshore, in the early Cretaceous, sands and clays were laid down in rivers and lakes. These Cretaceous deposits are found in small pockets in Cape Breton and central Nova Scotia, especially in the Shubenacadie and Musquodoboit Valley areas of the mainland. There is evidence that similar deposits may also occur in the Sussex area of New Brunswick. The Nova Scotian deposits consist largely of pure quartz sand, used in glass making, and of clays, some of which are used to make bricks. Among the clays is a white, aluminum-rich clay variety called "kaolinite". Kaolinite is ideal for making special types of pottery and for use in paper making, so there is a possibility that kaolinite from the Shubenacadie deposits will be commercially developed.

Kaolinite is formed by chemical breakdown of the mineral feldspar (common in granites of the South Mountain Batholith) in a wet, subtropical climate. So we can surmise that Nova Scotia's climate in the early Cretaceous was like parts of Central America today, and that it was wetter than in the Triassic or Jurassic. Other evidence for a warm, wet climate comes from onshore lignite deposits. Peat accumulated in lakes and is preserved as beds of lignite, the intermediate stage between peat and coal. Fossil wood associated with these deposits has growth rings, showing that the Cretaceous year in Nova Scotia had seasons. (Trees in nonseasonal climates, such as the modern tropics, do not develop growth rings.) However, these seasons may have been based on changing rainfall rather than on fluctuations in temperature.

The period of deposition of sand, clay and peat on a vast flood plain came to an end about 100 million years ago. This reflected the final split between Newfoundland and Labrador on one side and northern Europe on the other, thereby also splitting the ancient Avalon and Miramichi-Bras d'Or Terranes in two. The event affected areas as far south as the Maritimes. Earthquakes would have become more common as the land rose, and rivers started to cut deeply into the soft early Cretaceous sediments. More resistant rocks would have stood out as highlands, essentially in the same places that they do today.

Cretaceous Life—and Death

Climate change, herbivore feeding patterns and even the humble insects may have helped to trigger a major step in evolution: the appearance of flowering plants in the early Cretaceous. To human eyes, the pre-Cretaceous world would have seemed a relatively drab place without flowering plants, which include most deciduous trees. By the end of the Cretaceous, flowers had become the major competitors of the conifers. This change in vegetation seems to have happened alongside a change in dinosaur populations. The long-necked sauropods were replaced in the northern hemisphere by the armoured ankylosaurs, and by the ornithopods (or animals with "bird feet"); the latter group included the duck-billed dinosaurs and *Triceratops* and its relatives. Unfortunately, fossils of these exotic creatures are not found in the Maritimes, mainly because Cretaceous sedimentary rocks are rare onshore in our region. However, nonmarine rocks of Cretaceous age underlie the Scotian Shelf, and these undoubtedly contain fossils of the great reptiles.

The Cretaceous-Tertiary boundary marks a major extinction event. (The Tertiary extends from about 65 to 2 million years ago.) In the Maritimes, the only evidence for this boundary is in petroleum exploration wells offshore, in the Scotian Basin. The fossils here are mainly microfossils, some of which show the same extinction that decimated other plants and animals–including the dinosaurs. The question of why dinosaurs became extinct has fascinated scientists and non-scientists alike for many years. Many strange and now-discarded theories have been put forward. For example, one suggestion was that dinosaurs over-ate; another was that they ran out of food; and yet another was that small nocturnal mammals ate their eggs. It was even suggested that dinosaurs died out because of group senility. The biggest problem with most of these hypotheses is that, even if they were plausible for the extinction of the dinosaurs, they took no account of the simultaneous extinctions of other groups of organisms. An acceptable explanation for the Cretaceous-Tertiary extinction event must also account for the survival of certain groups. Why, for example, did dinosaurs become extinct but turtles and crocodiles survive?

So, what happened at the end of the Cretaceous? Today, the most common belief is that the collision (or "impact") of a meteorite or comet with Earth set off a chain of events that resulted in a major global catastrophe. The impact would have formed a major impact crater. It would also have caused a massive earthquake, a colossal tsunami and, more long term, a gas and ash cloud. Such a cloud would have encircled the globe much like that caused by the volcanic eruption of Tambora mentioned in Chapter 1—only on a much vaster scale. For organisms not directly in the path of the meteorite, the threat would have come from these after-effects.

If such a catastrophic collision occurred in the not-so-distant geological past, there should still be physical remains from it. And, indeed, in recent years, strong evidence for such a Cretaceous-Tertiary boundary impact event has been found. This evidence includes the chemistry of sedimentary rocks, the presence of shocked quartz crystals, and even the identification of the probable impact crater: the Chicxulub structure in Mexico.

As impressive as the evidence may be for an impact event, the most important testimony comes from the fossil record. Without such evidence, we would not know about dinosaurs, let alone that they became extinct. Can the fossils themselves provide any clues about whether the extinction occurred over only a few years, as suggested by the impact scenario, or over a much longer time span? For the dinosaurs, there is only one region where it is possible to examine their fossil record leading up to the Cretaceous-Tertiary boundary, and that's in the western interior of North America. At the time of writing, a recent exhaustive search for latest Cretaceous dinosaur bones in Montana and Alberta indicates no significant decline in dinosaurs leading up to their demise. Therefore, we can say that evidence from fossils is at least consistent with an abrupt extinction.

But some geologists remain unconvinced about the impact scenario. This is partly because, at the end of the Cretaceous, other unusual geological events were happening. There was a global lowering of sea level. And India

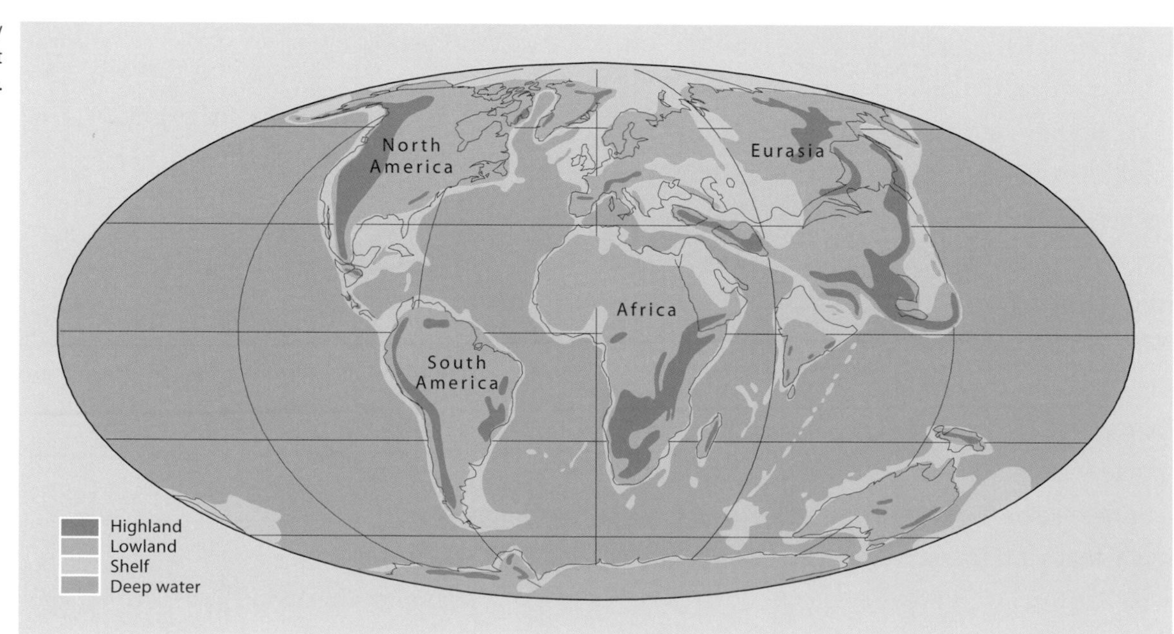

Global paleogeography of the early Cenozoic, about 50 million years ago.

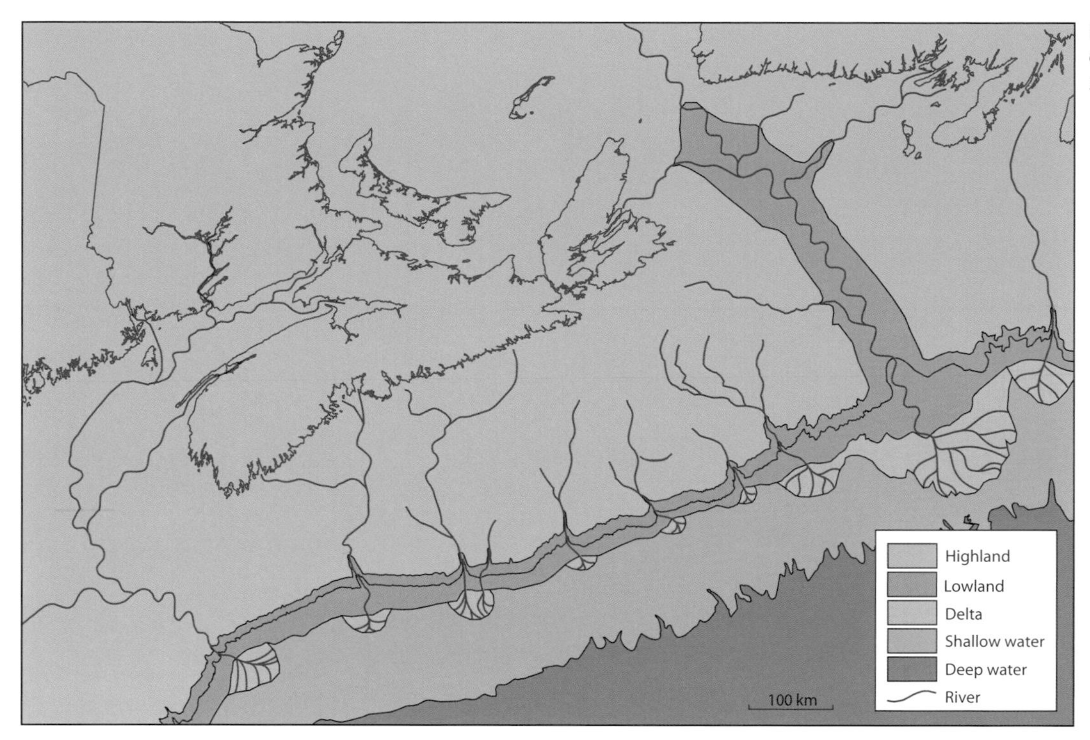

Regional paleogeography of the Maritimes about 30 million years ago.

was the site of one of the greater episodes of volcanic activity in geological history, which could have affected life in ways similar to an impact. Perhaps the fact that all these events occurred together within a short span of geological time provides the most reasonable explanation for the extinctions. One thing is certain: debate about the extinction of the dinosaurs and other animals about 65 million years ago will continue for many years.

The Age of Mammals

The Tertiary interval of the Cenozoic Era, comprising the older Paleogene Period and the younger Neogene Period, lasted from about 65 to 2 million years ago. During this time, the North Atlantic Ocean took on an increasingly modern outline. Most of the Maritimes was above sea level during the Tertiary. Erosion was therefore the dominant process, and the landscape was probably one of rounded mountains and hills, much as today.

With the dinosaurs extinct, the Tertiary became the Age of Mammals. A warm-blooded metabolism and the ability to give birth to live young make mammals highly adaptable, so that they rapidly populate land and sea. There are no Tertiary rocks on land in the Maritimes, so no Tertiary mammal fossils have been found here. However, other Tertiary fossils, mostly microfossils, are known from thick offshore sequences in the petroleum exploration wells. Some larger Tertiary fossils have been

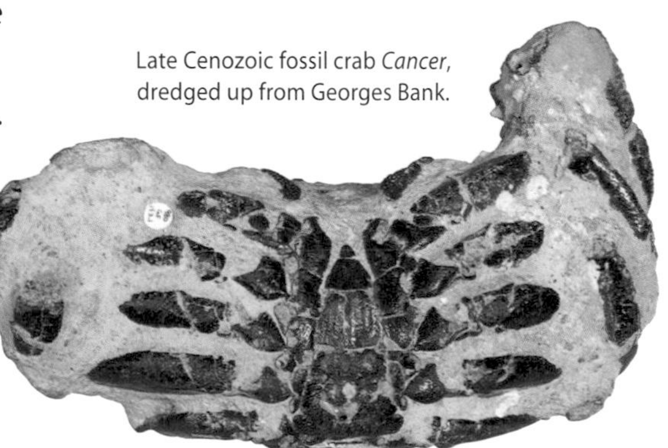

Late Cenozoic fossil crab *Cancer*, dredged up from Georges Bank.

found on the Scotian Shelf, dredged from the sea bottom by fishing trawlers .

The Cooling Trend

The Tertiary sediments of the Scotian Basin were deposited on a broad continental shelf, in an environment similar to that of today. A major change through the Tertiary, however, was the trend toward a cooler climate. One reason for cooling and increased seasonality in the Maritimes was the slow northward motion of this area on the now-separate North American plate. But changes were global, so other causes were involved.

Over time, sediment types slowly changed from a predominance of muds, with some limestone and chalk, to mainly sands. This was the result of a gradual fall in sea level as the global climate cooled and as more water was locked up in snow and icecaps. Therefore, conditions in our region changed from tropical in the Triassic to subtropical, temperate and, eventually, to arctic.

To find out why this dramatic cooling took place, we must continue our investigation into the changing patterns of continents and oceans, begun in Chapter 1. Reconstructions of these patterns show that, in the late Cretaceous, the distribution of oceans and continents resembled the greenhouse scenario, rather than the icehouse. There was some free flow of water around the Earth at low latitudes, but not in polar regions. This began to change about 60 million years ago, when southern continents were moving away from Antarctica and when northward migration of Australia was constricting the tropical flow of water between the Pacific and Indian oceans. At the same time, the tropical Tethys Ocean began to close. These plate migrations culminated in the opening of the Drake Passage south of Argentina, about 25 million years ago, and in the development of the Panama Isthmus, which closed the last equatorial oceanic "gateway" about 3 million years ago. The greenhouse world was thus transformed into an icehouse world.

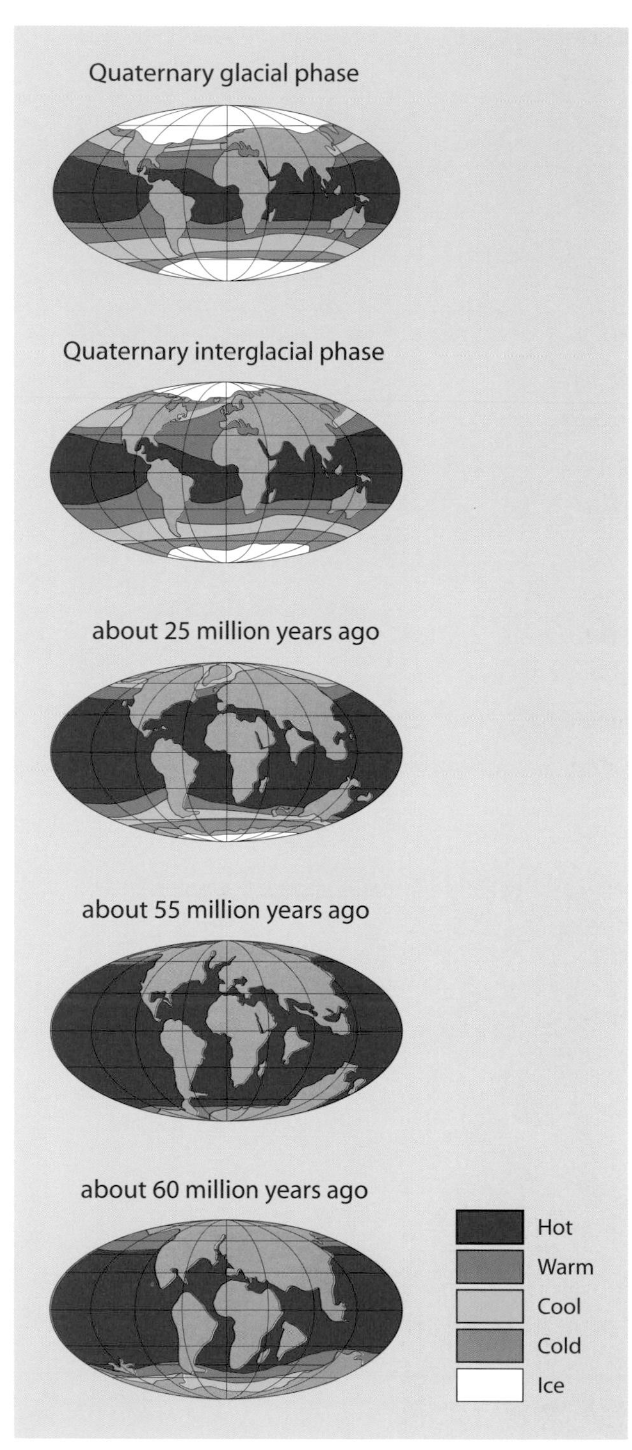

Sea temperature changes over the last 60 million years.

Antarctica was in a polar position for tens of millions of years before there was extensive ice. The first Antarctic glaciers at sea level appeared about 25 million years ago, at about the same time as the birth of the circum-Antarctic current. A widespread Antarctic ice cap did not form until perhaps 15 million years ago, because it was only then that wetter conditions provided enough snow. The presence of the Antarctic ice cap increased the amount of very cold water in the oceans and probably triggered modern deep oceanic circulation patterns.

Northern ice caps formed only 3 to 2 million years ago. As in Antarctica, the seemingly late onset probably had more to do with supply of moisture than with temperature. One theory is that the closure of the Panama Isthmus intensified the Gulf Stream, which carried warm water northward in the North Atlantic. This, in turn, produced snow in both northwestern Europe and northeastern North America, building up the ice sheets in these regions. Siberia, a region that epitomizes cold weather, did not develop an ice sheet. This was not because it was warmer there during the Ice Age, but because there wasn't enough moisture.

Hence, during the later Tertiary, glaciation became increasingly widespread. Similar cooling of the polar regions had occurred before, but during those times in the Ordovician-Silurian and Carboniferous, the Maritimes was near the equator and did not fully experience the glaciations. However, the effect of glaciation on the Maritimes in the Quaternary was extreme, as we shall discover in the next chapter.

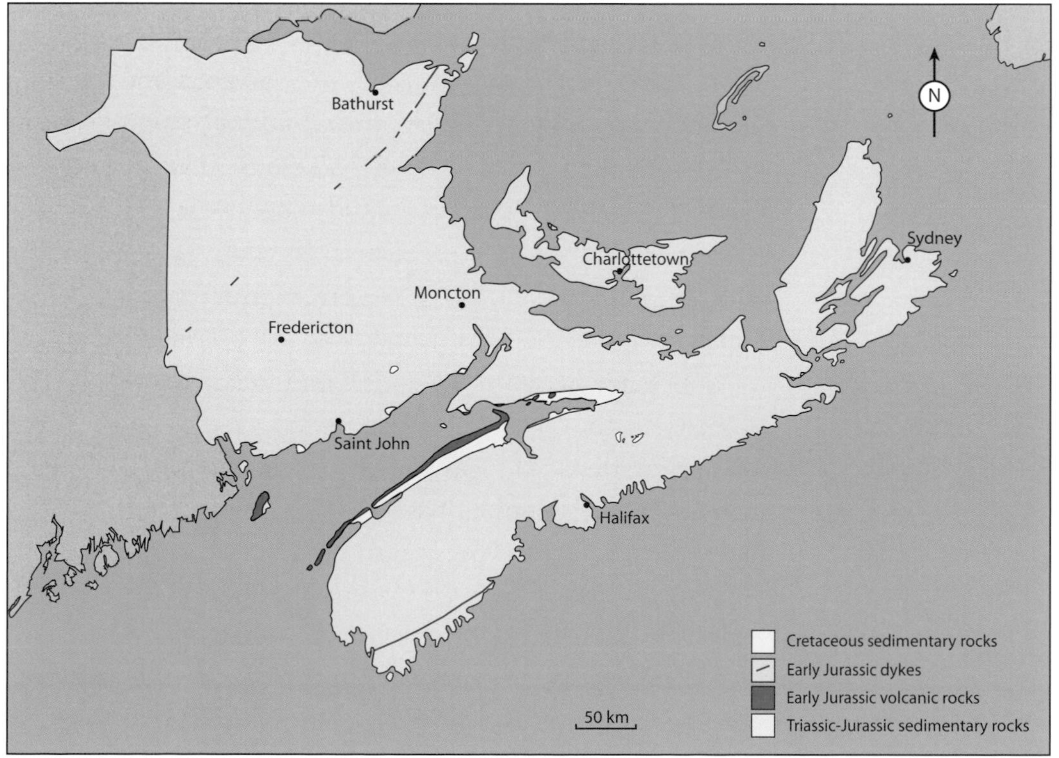

Distribution of Mesozoic rocks onshore in the Maritimes.

Museums

Museums of the Maritimes

There are a number of publicly and privately funded museums in the Maritimes that display geological material. An extensive listing of these is included at the back of the book. Here we highlight the three provincially operated museums that have geology as a main theme or major focus.

Fundy Geological Museum

Located on the Glooscap Trail in Parrsboro, the Fundy Geological Museum is operated by the Cumberland Geological Society as part of the Nova Scotia Museum. Opened in 1993, the Museum's collections, exhibits and interpretive programs are based on the geological resources of the Bay of Fundy and northern Nova Scotia. The museum highlights the area's fascinating paleontological and mineralogical treasures with displays, audio-visual presentations, interpretive programs, field trips and hands-on activities.

The Museum owes its origin in part to the wealth of fossil material found along the shores of the Bay of Fundy. At the Wasson Bluff site, located eight kilometres east of Parrsboro, a diverse assemblage of early Jurassic fossils was discovered in 1984. These fossils represent ancestral crocodiles, lizards, mammal-like reptiles, sharks, primitive fish and large and small dinosaurs (the oldest in Canada). Although many fossils from the site are fragmentary, parts of six prosauropod dinosaur skeletons have been found.

Visitors to the Museum can also travel 100 million years further back in time to the Coal Age forests by taking a forty-five-minute drive to Joggins. Over 140 years ago, two pioneering geologists, Sir William Dawson and Sir Charles Lyell, discovered the bones of some of the earliest reptiles inside fossil tree trunks. Such trunks can still be seen in the cliffs at Joggins, new ones being continually exposed as old ones are washed away by tides.

Scale model, at the Fundy Geological Museum, of the huge millepede-like animal, *Arthropleura*, crawling by fallen clubmoss tree trunks.

The Fundy Geological Museum sponsors the annual Nova Scotia Gem and Mineral Show, also known informally as the "Rockhound Roundup". And the Fundy Geological Museum displays several collections of the beautiful minerals to which "rockhounds" are attracted. These include amethysts, agates and many examples of zeolite minerals found in association with the North Mountain Basalt.

New Brunswick Museum

Skeleton of a mastodon, *Mammut americanum*, at the New Brunswick Museum. A similar mastodon was found in Quaternary deposits at Hillsborough, NB.

Late in April of 1996, the Prince of Wales officially opened the new exhibit facilities of the New Brunswick Museum at Market Square in Saint John. This modern facility houses fifteen hands-on, interactive galleries, which tell the story of New Brunswick and its place in the world. The new geology gallery houses several significant fossil displays, including the Hillsborough Mastodon, whose remains were discovered in 1936. Mastodons were large, woolly, elephant-relatives that lived in the Maritimes toward the end of the last glacial.

The Maritimes' oldest fossil, the Precambrian stromatolite *Archaeozoon acadiense*, is also on display at the New Brunswick Museum, along with Cambrian animals, Devonian plants and some beautiful fish skeletons and impressions. Some of the most attractive specimens are from "Fern Ledges", west of Saint John. These shale outcrops contain numerous plant fossils, amphibian tracks and insect remains.

Gesner's Museum, founded by Abraham Gesner in Saint John in 1842, was the first public museum in Canada and the first Natural History Museum in British North America. Three quarters of Gesner's 2,000 specimens were geological. Over the years, his collection has moved from one museum to another, survived the Great Saint John Fire in 1877, and now resides in the older part of the New Brunswick Museum, on Douglas Avenue in Saint John, where the collections are still housed. The natural history collections have grown considerably from Gesner's 2,000 specimens to over 300,000 and now form a superb resource for the people of New Brunswick.

Nova Scotia Museum of Natural History

In 1831, the Halifax Mechanic's Institute created a room for a museum. This museum flourished and, in 1868, by Act of Legislature, the Provincial Museum was established to include the Institute's collections. The museum moved many times until its present home on Summer Street in Halifax was built in 1970 to house the natural sciences and history collections. The Museum of Natural History is now part of a decentralized Nova Scotia Museum, a family of twenty-five separate museums throughout the province.

Of the 300,000 specimens in the Museum of Natural History collection, 55,000 are geological, comprising rocks, minerals and fossils—some of which were part of the Halifax Mechanic's Institute museum. Indeed, one specimen has been on display continuously since 1833. Displays at the museum, which emphasize the geology of Nova Scotia, lead one through the science of geology and the basics of how the Earth works. Visitors can discover what we learn from a rock, a formation, or the surface of our planet. Exhibit topics include time, plate tectonics, minerals, physical properties of minerals, rocks and their processes of formation, and fossils and their paleoenvironments.

The museum works closely with many institutions and other government departments, providing information, specimen loans and advice. Collection-based research at the Museum is an important contribution to our knowledge of the province's geology. The Museum also produces geologically-related publications and conducts extensive educational programs for schools and the general public. Other than the usual museum responsibilities of identifying specimens for members of the public and fielding other inquiries, educational activities include guided walks, lectures, school classes and workshops.

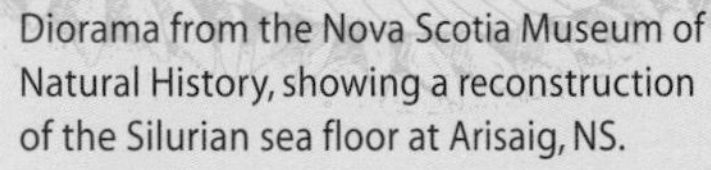

Diorama from the Nova Scotia Museum of Natural History, showing a reconstruction of the Silurian sea floor at Arisaig, NS.

A scene from the modern oceanic depths showing black smokers surrounded by a special fauna of tubeworms (right foreground), giant clams (left foreground) and spaghetti-like tubes of other worms (behind the clams). The scene is evocative of the kind of environment that produced the Ordovician mineral deposits in the Bathurst area of New Brunswick. The Ordovician fauna would have been different, but clams have been found in association with these mineral deposits.

CHAPTER 8
From Rocks to Riches

A Diversity of Wealth

From antiquity, people have used the Earth's natural resources to build and sustain civilization and to create a better place to live. Everything from basic necessities to sophisticated technology relies on raw materials from the Earth. And each house, office, school, hospital, factory and shopping centre is home to thousands of everyday items produced from rocks and minerals. Even a simple sheet of paper can contain several minerals. Indeed, there is a lot of truth to the saying "if it can't be grown, it has to be mined".

Aerial view of the gypsum quarry near Milford, NS

When asked about mining, most people think of metal ores, coal, or perhaps gypsum and salt. However, the range of resources that we extract from the Earth is much broader. Such geological resources also include ground water, petroleum, gemstones, sand, gravel, building stones and topsoil. In this chapter, we cover many of these materials. We discuss metallic resources in the sections "Iron—A Forgotten Heritage", "Black Smokers and Anoxic Oceans" (where we describe copper, lead and zinc sulphide deposits), "In Another Vein" (tin and related metals) and "A Golden Opportunity" (gold). Non-metallic resources include salt, gypsum and potash (discussed under "Nature's Evaporators"), building stone ("A Stone for All Seasons"), and aggregates such as sand and gravel ("True Grit"), and peat ("Vegetable or Mineral?").

Energy resources fuel our economy. They include coal ("A Revolutionary Fuel") and hydrocarbons such as crude oil and natural gas ("Energizing Our Economy"). Coal is an important energy source globally, although its popularity is waning because of its negative environmental impact. Historically, only minor quantities of oil and gas have been produced in the Maritimes. However, the region is now becoming a major producer, with the opening of several fields on the Scotian margin. Finally, we discuss ground water in a section entitled "Vital Fluid".

The topics discussed in this chapter demonstrate the diversity of geological resources in the Maritimes. And other resources, such as clays for bricks and pottery, would be included here if space weren't limited. Aspects that can't be omitted, however, involve the mining heritage passed down to us and the heritage that we will leave for future generations. Thus we include a section on our mining history, a fascinating story that is not widely known, and to close the chapter we consider issues of sustainable development and the environment.

Geological Background

Geological resources are found in all three rock types: igneous, sedimentary and metamorphic. Plutonic igneous rocks are the source of many metallic deposits. Tin, for example, is associated with granite, whereas nickel is associated with gabbro. Sedimentary rocks serve as reservoirs for oil, natural gas and ground water and may contain metal deposits. Some sediments and sedimentary rocks are a resource in themselves: examples are coal, gypsum, salt and some building stones. Metamorphic rocks provide us with marble, used as a building stone and in art.

Geological processes and structures are also important. Rifting at plate margins or in back-arc basins plays a vital role in the formation of metallic sulphide deposits, discussed later in this chapter. And folds and faults are crucial in forming oil, natural gas and ground-water reservoirs.

Resources form in numerous ways. A common thread for many geological resources is that they have been concentrated in some way. Gold, for example, is present in most rocks in minuscule amounts (1 to 3 parts per billion). Under certain conditions and over time, processes in the crust can selectively mobilize and transport gold, then concentrate it in particular rocks or structures. In this way, some rocks become so enriched in gold (5 to 30 or more parts per million) that it is worthwhile to mine them. Because the sand that we use in concrete and road construction is more common than gold, the need for a concentrating process is less obvious. But to be suitable, sand must have a certain grade or quality, and high-quality sand can be concentrated by the sorting action of wind or water currents.

Paleo-Indian arrowhead, about 500-1000 years old, from Enfield, NS.

Early Days

The native peoples pioneered the use of geological resources in the Maritimes. For example, they used rocks and minerals such as graphitic schist, hematite, limonite and agate to make paints for decorating skin, for clothing and tools, as well as for burial ceremonies. Native people used pyrite with chert to create sparks for starting fires, and they fashioned soft sedimentary and metamorphic rocks into elaborately crafted smoking pipes. Our knowledge of these items is fragmentary, however, since written records date back only a few hundred years.

From archeological studies, we know that the Mi'kmaq used ceramic vessels for both cooking and storage as early as 3,000 years ago. Clays found throughout the Maritimes were fashioned into coiled pots, which were often decorated and fired with crushed shells to make a durable finish. The Mi'kmaq also discovered ways of using the differing physical properties of such rocks as agate, slate and rhyolite to make tools through grinding and polishing. These tools included axes, adzes, mauls, hammers and weights for fish nets. The Mi'kmaq worked rock by a method known as flaking to create sharp implements such as scrapers, knives and projectile points.

Prehistoric sites in the Rustico area of PEI, and in Lunenburg County, NS, have yielded copper artifacts, including a tubular copper bead and various cutting implements. These finds indicate that the Mi'kmaq used copper before contact with Europeans and that they were possibly familiar with the use of heat for softening, annealing and even smelting this

Naturally occurring (or "native") copper from Cape d'Or, near Advocate Harbour, NS.

metal. The copper may have come from sites such as Cape d'Or, near Advocate Harbour, NS, where copper naturally occurs, or it may have been obtained through trade. Later, Europeans brought various copper utensils to North America, and these were commonly refashioned by the Mi'k-maq into a variety of other implements, including projectile points, cutting tools, beads, chisels, awls, fish hooks and pan pipes.

Early European explorers were motivated by the search for precious minerals, especially gold and silver. Sieur de Monts was one such pioneer who explored eastern Canada. In 1603, King Henry IV of France issued a charter to de Monts containing specific instructions "to make carefully to be sought and marked all sorts of mines of gold and silver, copper and other metals and minerals, to make them to be digged, drawn from the earth, purified and refined for to be converted into use ...." In 1604, de Monts' companion, Samuel de Champlain, set out from their encampment on St. Croix Island, NB, with a native guide to search for a copper "mine" reported the previous year. They didn't find the mine, but they did discover a small deposit of copper, probably at Beaver Harbour, NB.

In 1639, a small amount of coal was mined in the Minto area of New Brunswick and shipped by schooner to Boston. By 1686, Acadians quarried sandstone for grindstones at Isle aux Meulles (Grindstone Island) in Shepody Bay, south of Moncton, NB. And in 1701, French engineers rebuilding the fortifications of Port Royal, near modern Annapolis Royal, NS, used mortar made from limestone quarried at Musquash Harbour, near Saint John, NB.

Carboniferous rocks form Grindstone Island, in Shepody Bay, NB. Early Acadians quarried sandstone for grindstones on Grindstone Island (Isle aux Meulles).

In spite of these early developments, it was not until after the pioneering investigations of Abraham Gesner (1797-1864), James Robb (1815-1861) and Loring Bailey (1839-1925) that serious prospecting and mining efforts began in New Brunswick. Around 1848, iron and manganese began to be mined and smelted at Jacksonville near Woodstock. Within a few years, oil and natural gas were discovered at Saint-Joseph, gypsum was quarried at Hillsborough, coal was being mined continuously at Minto, and copper, lead and manganese were produced from a number of small mines near the Bay of Fundy. Albertite, a solid petroleum-like substance, was discovered in Albert County, southeastern NB, by two brothers named Duffy. Abraham Gesner staked the property but lost ownership of the mine after a long legal battle with men who insisted that albertite was coal. Gesner later developed a method of distilling liquid kerosene from albertite, however, and is now recognized as the father of the oil-refining industry because of his invention.

In Nova Scotia, coal was a major factor in attracting early Europeans to the region. Governor Nicholas Denys wrote in dispatches to France in 1673 that there was "a mountain of very good coal four leagues up Spanish River", near Cow Bay, Cape Breton Island, NS. By 1720, coal mines in the Glace Bay area were supplying the French fortress at Louisbourg. Acadians also began removing coal from the cliffs at Joggins. In Pictou County, coal was discovered on the East River in 1798 and was being mined by 1807.

Although coal was important to Cape Breton Island and northern Nova Scotia, it was the search for gold that helped open up the hinterland. The discovery that established the gold industry was made by Lieutenant L'Estrange in 1858, who found traces of gold in quartz veins near what is now Mooseland. In 1860, John Pulsiver of Musquodoboit and his guide, Joe Paul, discovered the same site, but found much more gold there. Several months later, Peter Mason, following a lead from Joe Paul, struck gold at the head of Tangier Harbour. These finds triggered Canada's first gold rush in the spring of 1861.

Iron—a Forgotten Heritage

Goethite, an iron ore mineral from Bridgeville, NS

Coal and gold continued to be important resources in the region through the twentieth century. Prior to the First World War, iron mining and smelting were also significant Maritime industries, but their role is nowadays almost forgotten. Early operations were small but important for providing tools for farms and local industry, as well as for homes and carriages.

The earliest discovery of iron deposits was in 1604, when Sieur de Monts reported finding veins of magnetite (magnetic iron oxide) in North Mountain Basalt near Digby, NS, and in the beach sands of St. Mary's Bay, NS. But commercial mining did not begin until 1825, when a few small mines near Torbrook and Bridgeville, NS, were developed. The ore from Torbrook was smelted near Clementsport, NS, and operations continued sporadically until 1916. The largest iron mining operation opened in 1849 at Londonderry, NS, and operated until the early twentieth century.

In all three areas, ore was smelted using small forges and blast furnaces, with iron production reaching a peak between the mid 1850s and mid 1870s. In 1874, at Londonderry, German inventor Friedrich Siemens (1823-1883) first tested his new open-hearth process for the direct conversion of molten iron into steel, a process which became one of the principal methods in steel making. Indeed, so significant were the iron mining and smelting operations at Londonderry that the settlement became, briefly, one of the largest in Nova Scotia.

A specimen of "Torbrook shell ore" from Torbrook, NS. The rock is made up of a precipitate of the iron oxide mineral, hematite, and of fossil brachiopod shells.

There are three types of iron ore in Nova Scotia: bedded, replacement vein, and residual. Bedded ores are sedimentary rocks formed by the precipitation of the mineral hematite (iron oxide) in shallow marine waters. The deposits at Torbrook are bedded ores of Devonian age. They are located in the Meguma Terrane, within a sequence that also includes shales and sandstones. Replacement vein deposits form when hot aqueous solutions from a nearby granite magma precipitate hematite and siderite (iron carbonate) in fractures in the rock, and also replace some of the surrounding rock. These deposits tend to be small but were mined to supply local needs.

By far the most important deposits have been residual ores found along a 22-kilometre belt in the Londonderry area. These deposits developed as the result of a natural two-stage process. A series of parallel, almost vertical lenses, each up to 30 metres wide, formed in fractures in the rock (but with no replacement of surrounding rock) during the intrusion of Devonian granites. These granites were intruded into volcanic and sedimentary rocks of the Cobequid Mountains. Originally, these lenses consisted mostly of the minerals siderite and ankerite (iron carbonates), but the lenses were subsequently leached by downward percolating ground water, which dissolved the carbonate and redeposited the iron as iron oxide (mostly as the minerals hematite and goethite). This process concentrated the iron from less than 10 percent in the original carbonate to nearly 60 percent in the ore.

Production of iron in Nova Scotia decreased gradually in the late nineteenth century with the discovery of large iron ore deposits in the Lake Superior region and at Wabana in Newfoundland. Nevertheless, steel continued to be made at Sydney, using iron ore from the Wabana mine.

One of the principal people involved in the operation of the Londonderry iron mines, George Drummond, also figured in the development of the Drummond Iron Mine at Austin Brook, south of Bathurst, NB. The Austin Brook iron deposit is a bedded ("stratiform") type formed in a volcanic setting by the venting of mineral-rich fluids onto the sea floor, as is described for other metal deposits in the next section. The mine opened in 1910, but closed in 1913 because the quality of the ore proved to be too poor. However, it was re-opened for a year during the Second World War, when it was operated by the Dominion Steel and Coal Company of Sydney, NS.

Bedded ironstone (iron sulphide) from an abandoned mine at Austin Brook, near Bathurst Mines, NB.

Black Smokers and Anoxic Oceans

In the Bathurst area of northern New Brunswick, numerous deposits of base metal (copper, lead and zinc) sulphides are associated with volcanic rocks. One of these, the Brunswick No.12 mine, is one of the largest zinc deposits in the world. The old Stirling Mine near Loch Lomond on Cape Breton Island, NS, is another example of a deposit of base and other metals associated with volcanic rocks. At Bathurst, the enclosing (or "host") strata are Ordovician volcanic and sedimentary rocks, whereas at Stirling, the rocks are Precambrian (about 680 million years old). In the Bathurst deposits, the minerals sphalerite (zinc sulphide), galena (lead sulphide), and chalcopyrite (copper-iron sulphide) occur together with pyrite (iron sulphide) in large bodies that lie parallel to the strata enclosing them. These "stratiform" ores are referred to as "massive sulphide" deposits, not because of their size but because they consist almost entirely of metallic sulphides.

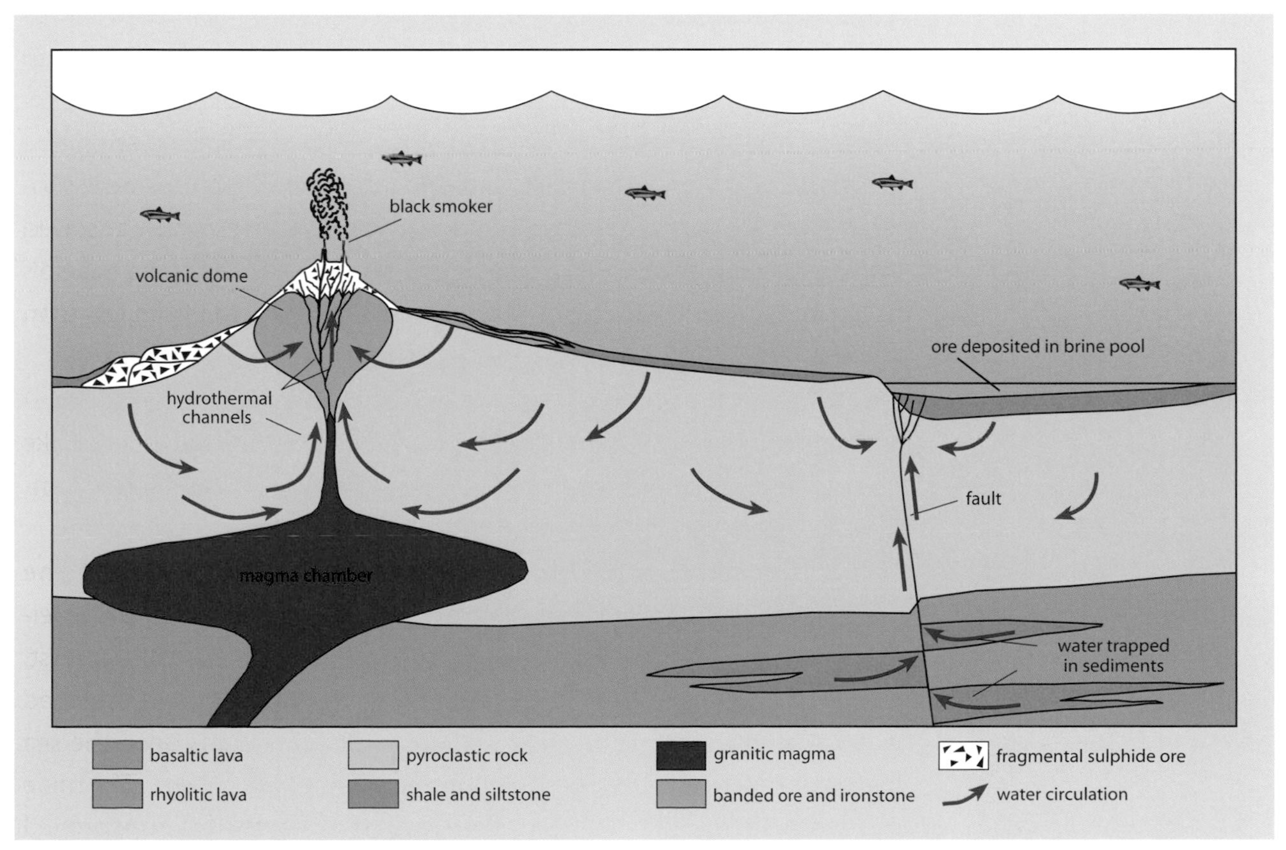

The formation of metalliferous massive sulphide deposits. Corrosive sea water seeps into the crust through fractures, heats up and leaches out the lead, zinc, copper and other metals present in rocks under the sea floor. The superheated, highly-pressured, mineral-laden water escapes back to the surface through fissures or faults, emerging as a black smoker. Under well oxygenated conditions, the leached metals disperse into the sea water. Under anoxic conditions, however, oxygen-rich sulphate, a normal component of sea water, converts to oxygen-poor sulphide, which readily combines with metals in hot water vented from black smokers to form the minerals pyrite, sphalerite, galena and chalcopyrite. These minerals are the major components of massive sulphide deposits, these deposits including both ore types listed in the legend.

Drill cores consisting of sedimentary and volcanic rocks deposited in the Tetagouche Back Arc Sea about 470 million years ago. Near Devil's Elbow, about 60 kilometres west of Bathurst, NB.

Since massive sulphide deposits are found not only in association with volcanic rocks, but also with marine sedimentary rocks, it is reasonable to assume that they must have accumulated on the sea floor. How this kind of deposit formed was not clear until deep-sea submersibles were used to study sea-floor volcanic "hot spots", such as those on the modern East Pacific Rise. At such hot spots, geologists were surprised to discover chimney-like structures, which they called "black smokers" (or "hydrothermal vents") emitting plumes of superhot water containing fine precipitates of metal sulphides. These metallic sulphides were deposited as layers and mounds around the black smokers. Even more surprising was the presence of thriving communities of bacteria, giant clams, tube worms and blind crabs living near the vents. These deep-sea communities differ from most life on Earth because they are not dependent on energy from the sun. Rather, bacteria in these communities break down hydrogen sulphide in the vents, in the process releasing energy. Other animals in the community live off the bacteria in various ways.

The first step in the formation of hydrothermal vents involves the seepage of sea water into the crust through fractures in volcanic rocks forming the sea floor. The water gradually becomes hotter as it seeps deeper because the source of the volcanic rocks is an underlying chamber or reservoir of magma. As the water becomes hotter, it expands and thus is under increasing pressure in its confined space. The water contains dissolved substances, including chlorine, that make it highly corrosive when hot, enabling it to leach out the lead, zinc, copper and other metals present in the volcanic rocks. When the superheated, highly pressured, mineral-laden water finds an escape route through a fissure or fault zone, it is forced back to the surface, emerging on the sea floor as a black smoker. On land, geysers such as Old Faithful at Yellowstone National Park in the United States function in a similar way.

Under conditions of normal oceanic circulation, metals accumulate initially as sulphides on the sea floor, but quickly become oxidized, their components diffusing into the sea water. So how do mineable deposits of metal-bearing sulphides become preserved? This question was answered when geologists realized that large volumes of water in an ocean basin can become deprived of oxygen. This "anoxia" results from poor circulation between the ocean's surface and bottom waters, and it prevents the decay of organic matter. Black shales rich in organic matter then accumulate on the sea floor. Such black shales were deposited widely in Ordovician oceans when the Bathurst ore bodies were being created. Anoxia also converts oxygen-rich sulphate, a normal component of sea water, to oxygen-poor sulphide, which readily combines with metals in hot water vented from black smokers to form the minerals pyrite, sphalerite, galena and chalcopyrite.

Gold-bearing quartz veins from The Ovens, NS.

In Another Vein

Granitic rocks underlie large areas of the Maritime Provinces, including much of southwestern and central New Brunswick, as well as southern Nova Scotia where the South Mountain Batholith extends from Halifax to Yarmouth. These Devonian granites were formed as the Rheic Ocean closed and its opposing continental margins collided, forcing crustal rocks upward into mountains and downward into hotter regions of the Earth's interior. Eventually, melting of the lower crust produced huge bodies of granitic magma that were more buoyant than the surrounding, denser rocks. As described in Chapter 5, these granitic bodies "floated" upward into the core of the mountains, where they cooled and solidified. Over time, the mountains have been eroded and the granitic rocks exposed.

As granitic magma moves upward, it cools and crystals form. This process of crystallization selectively removes some substances from the magma, thus enriching the concentration of the remaining substances. These enriched substances include metallic elements (such as copper, zinc, tin, tungsten and gold) and volatile elements and compounds (such as water, carbon dioxide, fluorine and chlorine). As more crystals form, metal- and volatile-rich fluids accumulate near the roof of the shrinking magma chamber. In the final stages of crystallization, the volatile elements and compounds have so much less space in the greatly reduced magma chamber that they exert sufficient pressure to fracture both the granite and the surrounding rock. The metal- and volatile-rich fluid then flows into the fractures and solidifies to form veins enriched with metallic minerals.

Metallic mineral concentrations formed in this way may be economically valuable. In New Brunswick, tungsten ores at Burnthill Brook, about 70 kilometres due north of Fredericton on the Southwest Miramichi River, antimony ores at Lake George, near Fredericton, and zinc-lead-silver-copper ores at Nigadoo, near Bathurst, are examples of vein-type deposits that formed outside the parent granite. All of these deposits supported mining operations in the past. Semi-precious gems, such as topaz, beryl, tourmaline and fluorite, may also occur in granite-related veins.

Copper at Woodstock, NB, tin at East Kemptville near Yarmouth, NS, and the Mount Pleasant tin-tungsten-molybdenum-copper-bismuth-indium deposit near St. George, NB, are examples of granite-related deposits in which the metals are disseminated within the parent granitic rocks.

Gold in quartz, from Montague Gold Mines, NS.

Still other granite-related deposits are termed "skarns", and typically occur in limestones or lime-rich sedimentary rocks adjacent to granite intrusions. Skarns are named for a characteristic suite of minerals formed from the chemical reaction between fluids coming from the granite and carbonate minerals in the host rock. In northern New Brunswick, the Popelogan, Patapedia and MacKenzie Gulch copper deposits are examples of skarn-type mineralization.

A Golden Opportunity

Gold in Nova Scotia occurs in the 15-kilometre-thick Cambrian to Ordovician turbidite strata of the Meguma Terrane. After the turbidite beds were deposited offshore, they were compressed into large folds as the Rheic Ocean closed during the Silurian and Devonian. Each of these folds is tens of kilometres long and runs parallel to the long axis of mainland Nova Scotia. The gold occurs as microscopic to nugget-sized particles within or next to narrow quartz veins found near the crests of the folds. The common occurrence of the gold-coloured mineral pyrite ("fool's gold") in the same rocks can confuse the unwary.

To understand how gold came to be in the quartz veins, we need to know how the quartz veins themselves were formed–a subject of considerable debate. It has been uncertain when the veins formed and whether the gold originated within the Meguma rocks or deep in the crust. However, one theory now emerging is based on studies of several deposits (Beaver Dam, Caribou, Forest Hill, West Gore and Moose River) in the eastern shore region of Nova Scotia. These deposits were explored and developed during the 1980s. The theory also explains deposits at The Ovens, NS, where small amounts of gold in bedrock and as placer deposits in beach sand have been produced in the past. (Placer deposits are explained later in this section.)

According to this theory, as the Rheic Ocean closed, the collision of continental margins folded the sedimentary rocks of the Meguma Terrane. With continuing collision, the folds became ever tighter, causing slippage along bedding surfaces. For a better idea of this process, fold a telephone directory. At first, the book folds as a unit, but as you fold more tightly, the pages begin to slip against one another. These bedding surfaces ("between the pages") were zones where fluids could migrate and accumulate.

The fluids may have been generated by the same process deep in the crust that was forming the granitic magma. As they forced apart the bedding surfaces, the fluids created spaces in which dissolved material started to form minerals such as quartz, gold, and sulphides such as pyrite. The presence of visible gold suggests that, over time, as the fluids continued to circulate, they re-dissolved, transported and reprecipitated the precious metal, concentrating it in the process.

Caves at The Ovens, NS, a gold-mining locality. At least one of the caves was enlarged during mining.

Although most veins are found on the sides (or "limbs") of the folds, in some cases they are at the crest (or "hinge"). In our analogy, a tightly-folded telephone directory would develop spaces in the hinge region, suggesting that fold crests in rocks would be ideal places for fluids to accumulate. This is precisely what happened in some of the Meguma deposits, the most notable example being the Salmon River deposit at Port Dufferin, NS.

In New Brunswick, most of the important gold deposits are related to Ordovician to Devonian granites or to major fault zones. As mentioned in the previous section, metals such as gold may be concentrated as magma cools and crystallizes. The gold may then precipitate in mineralized veins. At Clarence Stream, northeast of St. Stephen, gold is present in veins formed deep in the crust and later exposed by uplift and erosion. In the Annidale and Shannon areas west of Sussex, and at Poplar Mountain southwest of Woodstock, gold seems to have been deposited in veins and altered rock closer to the surface.

Many gold-bearing veins are associated with major fault zones. In such settings, the gold is probably formed deep in the crust through leaching of rocks by hot circulating fluids, which then migrate to favourable structural sites, such as faults. One example is the gold mineralization near Cape Spencer, southeast of Saint John, NB.

Ancient hot-spring activity in Carboniferous sedimentary rocks in southeastern New Brunswick may also have caused important gold mineralization. Hot-spring mineralization is similar to base-metal mineralization formed around black smokers on the sea floor. Hot-spring mineralization, however, occurs mainly on land.

Mineral deposits exposed at the surface are subject to erosion, just like other rocks. Transportation and deposition of these eroded materials, mainly in streams, tend to concentrate the heavier particles, including gold. Economic deposits formed in this way are called "placers". Gold in placer deposits occurs as nuggets, grains and dust, all mixed with gravel, sand and silt in stream valleys or along beaches. Panning is one method of extracting gold from such deposits. Klondike gold, from the Yukon, is placer gold.

Potash mine at Cassidy Lake, south of Sussex, NB.

Placer deposits can be buried and transformed into rock, and later re-exposed by erosion. A small gold occurrence in Late Carboniferous sedimentary rocks near Memramcook, NB, may represent such a "paleoplacer".

Nature's Evaporators

The thick potash beds near Sussex, NB, the salt at Pugwash, NS, and the widespread gypsum beds of central Nova Scotia—all of Carboniferous age—were precipitated from sea water through evaporation. Many dissolved salts derived from the erosion of rocks on land are carried by rivers and streams into lakes and seas. Under arid conditions, these bodies of water can dry up completely, reprecipitating the salt. During evaporation, dissolved salts are progressively concentrated until certain minerals begin to form crystals. First to crystallize is calcite (calcium carbonate), which settles to the bottom and forms limestone beds, followed by gypsum (calcium sulphate), then rock salt , or "halite" (sodium chloride), and, finally, potash (potassium chloride and related potassium compounds). Thus a complete cycle of evaporation forms a sequence of limestone, gypsum, salt and potash.

During the early Carboniferous, when the Maritimes was near the equator and rainfall was uncommon, evaporite deposits up to 500 metres thick were precipitated from the shallow Windsor Sea. This is a much greater thickness than can be accounted for by the evaporation of sea water from a single inland sea. Instead, the deposits must have originated from evaporation in restricted bays that were cut off from the sea but periodically replenished with sea water. In the dry climate, salts were formed by evaporation of this sea water over a long period of time. A similar, though artificial, procedure of extracting salt is followed today in many parts of the world. In these places, sea water is pumped continually onto shallow flats ("salinas") and allowed to evaporate.

Limestone and gypsum beds formed through evaporation underlie major parts of central Nova Scotia, although much of the salt deposited in the Windsor Sea has been dissolved by ground water. Also associated with the Windsor Sea are coral reefs. Perhaps these were barrier reefs, which isolated the sea water in lagoon-like bays. At some time after burial, the limestone, gypsum and reefs were mineralized by circulating waters rich in salts and charged with metals such as lead and zinc. The dissolved metals were precipitated when they encountered the calcareous beds and reefs to form pods and lenses of sphalerite (zinc sulphide) and galena (lead sulphide). Until recently, these ores were mined at Gays River, NS.

A Stone for Every Season

"Building stone" is any rock that can be collected or quarried and prepared for construction or ornamental use. It can be humble field stone used to build simple rubble walls or the finest granite, fit for a monument or the walls of a church. For such major architectural projects, commercial operations mechanically excavate immense slabs of rock from a quarry and shape them into blocks or sheets.

Field stone wall of various local early Paleozoic metasediments from the Meguma Terrane. Cape St. Mary, NS.

The Maritimes possesses many attractive buildings constructed of stone quarried within the three provinces, although some outside stone has also been used. Sandstone and granite are by far the most common types of building stone quarried in the region. As outlined in Chapter 1, sandstone is a sedimentary rock composed essentially of non-interlocking mineral grains held together by a cementing matrix. In granite, however, the mineral grains are fused and interlocking. Sandstone is therefore easier and cheaper to quarry and shape than granite, but it is more susceptible to erosion.

The MacDonald Farmhouse was built around 1820 at Bartibog, NB. It is one of the oldest stone houses in New Brunswick and is a Provincial Historical Site. As with many pre-1840 stone houses, the rock came from the nearest suitable outcrop, in this case a Carboniferous sandstone found less than a kilometre away.

Almost all sandstone quarried in the Maritimes is of Carboniferous to Permian age. Some quarries have produced building stone with tints of red, brown or purple, whereas others are grey, olive, or buff. In New Brunswick, the most significant sandstone operations are found in or near Miramichi, Chaleur Bay, Shediac, Sackville and Shepody Bay.

Stone from the Shepody Bay quarries appears in several Fredericton, NB, buildings, such as Christ Church Cathedral and the Legislative Building. Mount Allison University in Sackville, NB, has both red and olive sandstone buildings. The red sandstone is derived locally and the olive sandstone comes from a quarry at Rockport, NB. Prince William Street in Saint John, NB, has an impressive concentration of stone buildings, with material from about a dozen separate provincial quarries. The original New Brunswick Museum building in Saint John features stone from two provinces: its olive sandstone walls originated in Shediac, NB, and its restored sandstone steps are from Wallace, NS.

Wallace sandstone has been used also for two of the Maritimes' most famous buildings. One is Province House in Halifax. The other is Province House in Charlottetown, PEI, the site of the 1864 conference that led to

Detail of the Owens Art Gallery at Sackville, NB. The building was completed in 1895 and is constructed mainly of Carboniferous sandstone from a quarry at Rockport, NB. The uniformly fine-grained texture of the sandstone allows it to be carved into intricate structures, as shown here.

Canadian confederation. In Nova Scotia, sandstone has also been quarried at Amherst and Pictou. The old town hall and the police station in Amherst, for example, are made of locally quarried red sandstone. Similar sandstones have been quarried on Prince Edward Island, where they are referred to as "Island stones". Examples of buildings constructed from Island stones include St. Paul's Church and All Souls Chapel, both in Charlottetown.

The Maritimes, then, has over two hundred public and private buildings that reflect the beauty and diversity of our local sandstone. Most were built between about 1810 and 1920 when stone was relatively inexpensive to work and fashionable to use. Maritime sandstone was also coveted by American and central Canadian builders. Much was exported to places such as Boston and Philadelphia, as well as New York, where it can be seen still in some of that city's famous brownstone houses and the bridges of Central Park. Several of the parliament buildings in Ottawa and the Ontario legislature in Toronto also contain New Brunswick sandstone.

Granite, although expensive, has always been a popular building stone because of its beauty and durability. It is most commonly used where long life is essential—for bridges, walls, foundations, steps, pillars and curbing. Examples are the Grand Parade wall in Halifax, NS, and the walls and steps of Acadia's University Hall in Wolfville, NS.

There are several sources of granite in the Maritime Provinces. Probably the longest continuously operating granite quarries in the region are those at Hampstead, north of Saint John, NB. Although Hampstead granite was originally used to construct entire buildings—such as the Kings County Jail in Hampton, NB—the stone is used today primarily as monument bases, curbing and patio stones.

The red, pink, grey and black plutonic rocks (granites to gabbros) quarried at St. George, NB, were worked from the late nineteenth to mid-twentieth centuries. Stone from these quarries was used mainly in pillars and monuments, for which it is ideally suited. It takes a high polish that, even after many decades, still gleams as though it had just left the polishing mill. Many major towns and cities in New England, Ontario and the Maritimes have older buildings or monuments decorated by pillars or columns of St. George granite.

Nova Scotia also had several important granite quarries, including ones near Shelburne, Halifax and Guysborough. Today, however, a small amount of granite is produced only at Lawrencetown (Annapolis County) and in the Nictaux area. Perhaps the most impressive granite building in the Maritimes is St. Mary's Basilica in Halifax, NS, which is topped by a 45-metre spire—the tallest granite steeple in North America. The facade of St. Mary's is constructed of granite from the South Mountain Batholith, with minor decoration consisting of red granite from Scotland.

Historic Properties in Halifax, NS, comprises several stone buildings dating from the early to middle nineteenth century. They are built of local "ironstone" (a fine-grained metasedimentary rock from the Meguma Terrane) with granite and sandstone trim.

Maritime building stone other than sandstone and granite includes limestone, quartzite and slate. The region has few limestone deposits suitable for commercial building stone, so the limestone used for Acadia University's War Memorial Gymnasium at Wolfville, NS, is from the famous Tyndall stone quarries in southern Manitoba. Tyndall stone is a fossiliferous limestone of late Ordovician age teeming with gastropods, corals and brachiopods.

Metamorphic rocks are sometimes used as building stones. For example, a small deposit of quartzite (metamorphosed sandstone) was quarried at White Rock, NS, for use in some local buildings. Over a century ago, slate for the roof of the old Bathurst court house was quarried from the Tetagouche River area in northern New Brunswick. The Precambrian marble used to build the Cathedral of Immaculate Conception and Trinity Church in Saint John, NB, was quarried from nearby Rothesay and Indiantown. The pure white marble seen in some Maritime building facades, monuments and other settings has almost always been imported from Italy or Vermont. Generally, metamorphic rocks are too fractured or otherwise unsuitable as building stone. They are better put to other, related uses. For example, material from Nova Scotia's Marble Mountain quarry is crushed into ornamental landscaping chips, and is thus more a form of aggregate (see "True Grit", below) than building stone.

Gravestones are also made of "building stone". These gravestones are in the Old Burying Ground in Halifax, NS.

Graveyards also can be geologically fascinating places. It is possible, for example, to estimate the age of many Maritime graveyards from the type of stones used. Early European settlers lacked the tools or time to carve elaborate memorials, so they usually created their gravestones from sandstone or slate, which are easier to work. Around the 1850s, Italian marble headstones began to be used for the graves of wealthier citizens. In the 1870s, granite from St. George and elsewhere became available, roughly coinciding with improvements in granite-etching techniques. Today, granite remains the tombstone material of choice (or obligation, at some sites) as it is the most durable stone, and comes in a range of colours. Acid precipitation in recent decades has obliterated engravings on many expensive marble gravestones, whereas tributes on older, more humble but more durable sandstone monuments are still quite visible.

Gravel pit in the glacial stream deposits of an esker (Chapter 9), near Sixth Lake in the McAdam area of New Brunswick.

True Grit

Few think of aggregates–sand, gravel and quarried stone–as important geological resources. Yet these are basic materials for construction. They are used alone as ballast or fill or, more commonly, are mixed with a cement to produce concrete and mortar, or asphalt for road paving. Most aggregates are used for structural purposes such as highways, buildings and bridges. Less obvious uses include water filtration systems, traction sand and erosion control. Pure silica sand such as that found in Cretaceous deposits at Shubenacadie, NS, can be used in glass-making.

The two basic types of natural aggregate materials are crushed stone and granular aggregate. (Slag, a waste product of steel production also used as an aggregate, is not a natural material.) Granular aggregate is found as sand and gravel in unconsolidated surficial deposits. Early European settlers removed beach sand and gravel for their aggregate needs. During the nineteenth century, it was discovered that many of the islands in Mahone Bay, NS, had excellent beach-sand deposits. A flourishing trade developed, with sand loaded onto schooners by hand and shipped to Halifax to make concrete. Today, beach materials are no longer used because of environmental concerns, but sea shells seen in some early concrete structures serve as a reminder of their former use.

The dominant sources of sand and gravel today are meltwater deposits associated with glaciers. These deposits are distinct from other glacial deposits, such as till, because they have been washed and sorted in water. Meltwater deposits formed either in direct contact with the ice, such as the deposits of eskers, or in front of the glacier as outwash plains deposited in glacial streams. The quality of sand and gravel in these deposits depends mainly on the original rock source of the sediment. Igneous and metamorphic source rocks produce better aggregates than softer sedimentary rocks because they are more durable and yield a material with less silt and clay.

Sand and gravel quarry in the Parrsboro Valley, near Parrsboro, NS. The material being quarried is from an isolated hill called a "kame". Sand and gravel transported by flowing water on top of the valley glacier collected in a depression in the ice. When the glacier melted, the sediment-filled depression became a small isolated hill, quarried today as a source of aggregate.

Other sources of granular aggregate include modern stream and river deposits, but these are rarely used today because they tend to be small and because of environmental concerns. Weathered bedrock (especially granite) and some gravelly tills have been used as aggregate where low-quality material is acceptable or better material unavailable. Extensive gravel and sand deposits are also found offshore. These deposits are as yet untapped, but much of this material is within reach of modern dredging equipment. The offshore sediments originated as glacial till or outwash on land, the deposits being later cleaned and sorted by strong wave action as sea level rose and submerged them.

Many communities in the Maritimes use crushed stone for aggregate. Indeed, crushed stone, rather than granular deposits, is the primary source of aggregate, because of its generally higher quality and the lack of granular materials in parts of our region. Many construction projects now require crushed stone as part of their specifications.

Generally, the harder the rock, the better the aggregate. Rocks in the Maritimes older than late Devonian have generally been hardened by metamorphism and so make excellent aggregate. Metasediments of the Meguma Terrane, for example, are widely quarried in southern Nova Scotia for this purpose. However, other metamorphic rocks (such as schists and slates) are weak because of cleavage, making them susceptible to water damage and thus poor sources of aggregate.

Igneous rocks–such as the granite of the South Mountain Batholith in Nova Scotia and the Triassic basalts around the Bay of Fundy–have been used for many years as aggregates. One of the best examples of a modern aggregate quarry is in igneous rocks at Cape Porcupine, near Aulds Cove on the Nova Scotia mainland side of the Strait of Canso. This quarry was developed in the 1950s to supply fill for the Canso Causeway. Since completion of the causeway, the quarry has continued to operate, and now ships more than a million tonnes of high quality stone to other parts of eastern North America each year.

Vegetable or Mineral?

The peat found today in the Maritimes started to form about 10,000 years ago, shortly after the glaciers melted. In shallow, water-filled depressions, where stagnant conditions prevail, the lack of oxygen slows the activity of decay-causing micro-organisms. When the rate of plant production is faster than the rate of decomposition, plant remains gradually accumulate to form thick deposits of peat in what are called "peatlands". Peatland is a type of organic wetland that is often dome-shaped, forming a "raised bog". The surface is dominated by *Sphagnum* moss and shrubs of the heather family, lichens and sedges.

Peat forms cliffs several metres high at Escuminac, NB.

Peat deposits were traditionally used by the ancestors of many Maritimers as a local fuel supply. This example of peat cuttings is on the Isle of Skye, Scotland.

Peat is mined in all three Maritime Provinces, but principally in New Brunswick, which is the leading peat-producing province in Canada. Although peat has been used in Europe as a source of heat for hundreds of years, its main use today is as a soil conditioner. Peat alone has no nutritive value for plants, but when combined with fertilizers it produces one of the best growing media available to greenhouse growers. Peat is also used to treat waste water and to clean up oil spills.

A Revolutionary Fuel

The popularity of coal is waning because of the impact of its combustion by-products on the environment, especially carbon dioxide, which may contribute to global warming. Nevertheless, coal has been mined since prehistoric times, fueled the Industrial Revolution, and continues to be a major energy source globally. Over half of the electricity in North America is produced from coal, and construction of new coal-fired plants is outpacing construction of any other form of power generation plant.

Aerial view of peat workings, Baie Sainte Anne, NB.

The Maritime Provinces contain several large coal basins, all of Carboniferous age. The coal was formed about 310 million years ago, when our region was much closer to the equator and the climate was generally hot and humid. Luxuriant forests covered large lowland areas that were periodically flooded. Such flooding killed the vegetation, which accumulated to form peat, much in the way described in the previous section, although formed from different plants. The story of the formation of coal, the driving force of the industrial revolution, is told in Chapter 6.

Coal seam with fossil soil beneath it. Point Aconi, NS.

Coalfields are widespread in New Brunswick and Nova Scotia. The wealth of Nova Scotia in the nineteenth century and part of the twentieth century closely parallelled the growth of the coal industry, especially in Cape Breton. Coal mining in the Sydney Basin has been continuous since Governor DesBarres of the Colony of Cape Breton opened a mine in 1785. By 1870, the number of collieries had risen to 20 and continued to increase until the 1940s, when production peaked. Other areas of Nova Scotia with a long history of mining are Inverness and Port Hood in western Cape Breton, around Pictou, and Cumberland County. Mining began in Pictou in 1813 and ended with the Westray tragedy in 1992. In Cumberland County, several mining communities such as Joggins and River Hebert became important centres, the last colliery closing down in 1980. The

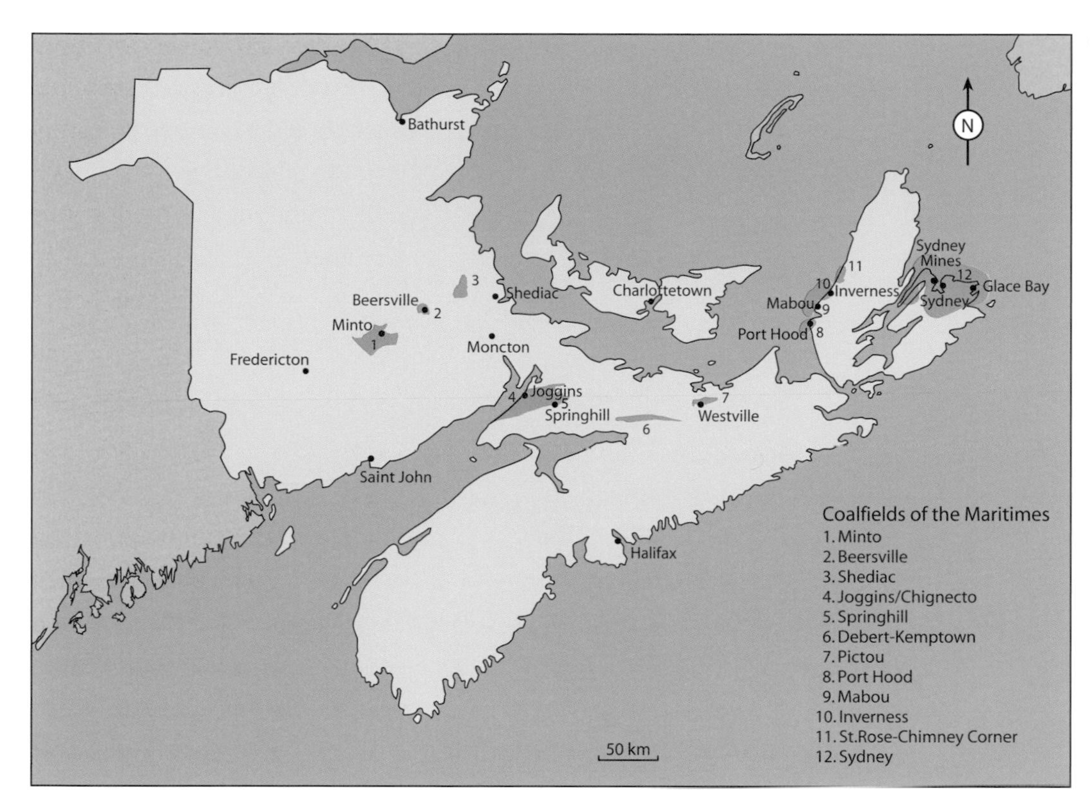

Coalfields of the Maritimes.

major production in the county, however, was from the Springhill mines. These shut down in the 1950s after major disasters in 1956 and 1958.

In New Brunswick, coal has been mined in the Minto area for many years, first by underground methods and, more recently, by "strip mining". In this process, enormous excavators called "draglines" remove the soil and rock covering the coal, which is then removed by smaller equipment.

The Carboniferous coals below the surface of parts of all three Maritime Provinces and much of the southern Gulf of St. Lawrence are a potential source of natural gas. At high temperatures, the complex organic molecules in deeply buried coal break apart (or "crack") to produce simpler molecules—mainly the natural gas, methane. This "coal-bed methane" stays trapped within the coal and can be dangerous if the coal is mined, sometimes causing fires or explosions. But methane is a valuable resource. It is one of the cleanest and most energy-efficient of all fossil fuels. Technology already exists to convert mined coal to natural gas and

Tailings from past coal-mining activity at Inverness, NS.

to other fuels such as methanol, gasoline, diesel and hydrogen. And new technology is opening up the possibility of extracting methane from coal by drilling, so it could become a major energy source in future years.

Energizing Our Economy

Petroleum, a word meaning "rock oil", includes natural gas, condensates, liquid oils (crude oil), and solids (bitumen). Condensate is a gas when underground, because of higher temperature there, but condenses to a liquid during its rise to the surface. Petroleum is generally found in sedimentary rocks, as in the Carboniferous of southern New Brunswick and in the Jurassic and Cretaceous of the Scotian Basin, off Nova Scotia.

Most scientists believe that petroleum is formed from decomposing animal and plant (organic) matter in an oxygen-poor (anoxic) environment. After burial, over millions of years, this organic material changes to petroleum. Petroleum is a mixture of hydrocarbon compounds, so-called because they all contain hydrogen and carbon. However, they usually also contain oxygen, nitrogen and sulphur. The hydrocarbon compounds in natural gas have a simple structure, whereas those in crude oil are complex, and those in bitumen are extremely complex.

Since large-scale production of organic material only became possible about two billion years ago with the evolution of eukaryotic organisms, very ancient rocks could produce only negligible amounts of petroleum. Until about 430 million years ago, there were no land plants, so all organic material was produced by marine animals and plants. Today, land plants are an important source of the organic matter in sediments, although their contribution is still less than that of marine organisms. The type of organic material in the rock is important because it influences the kind of petroleum produced. Marine plankton contain high percentages of lipids (fats)

How water, oil and gas occur between grains in a sedimentary rock such as sandstone. A) Water fills the spaces. B) Oil fills the spaces but a thin film of water surrounds each grain. C) Gas fills the spaces but a thin film of water surrounds each grain.

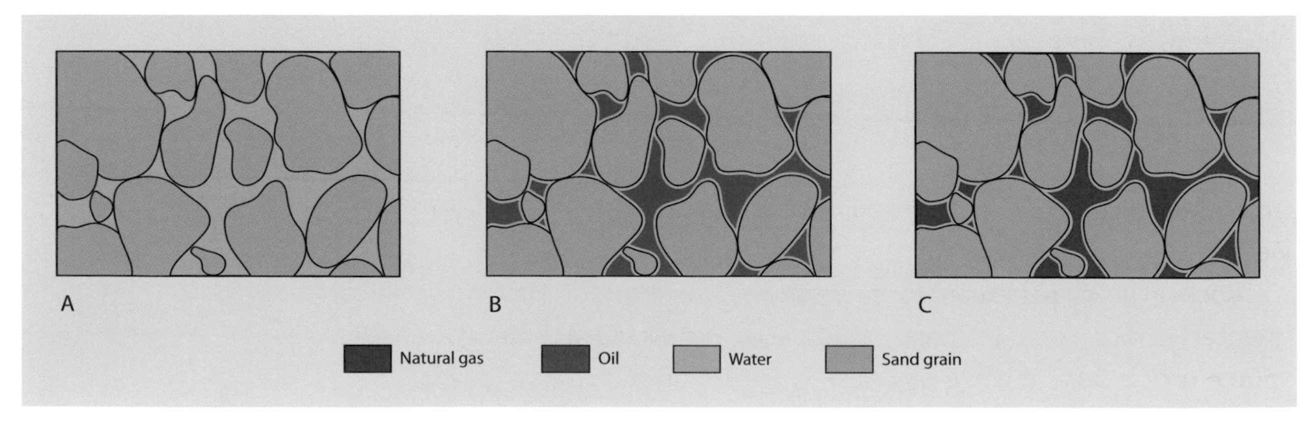

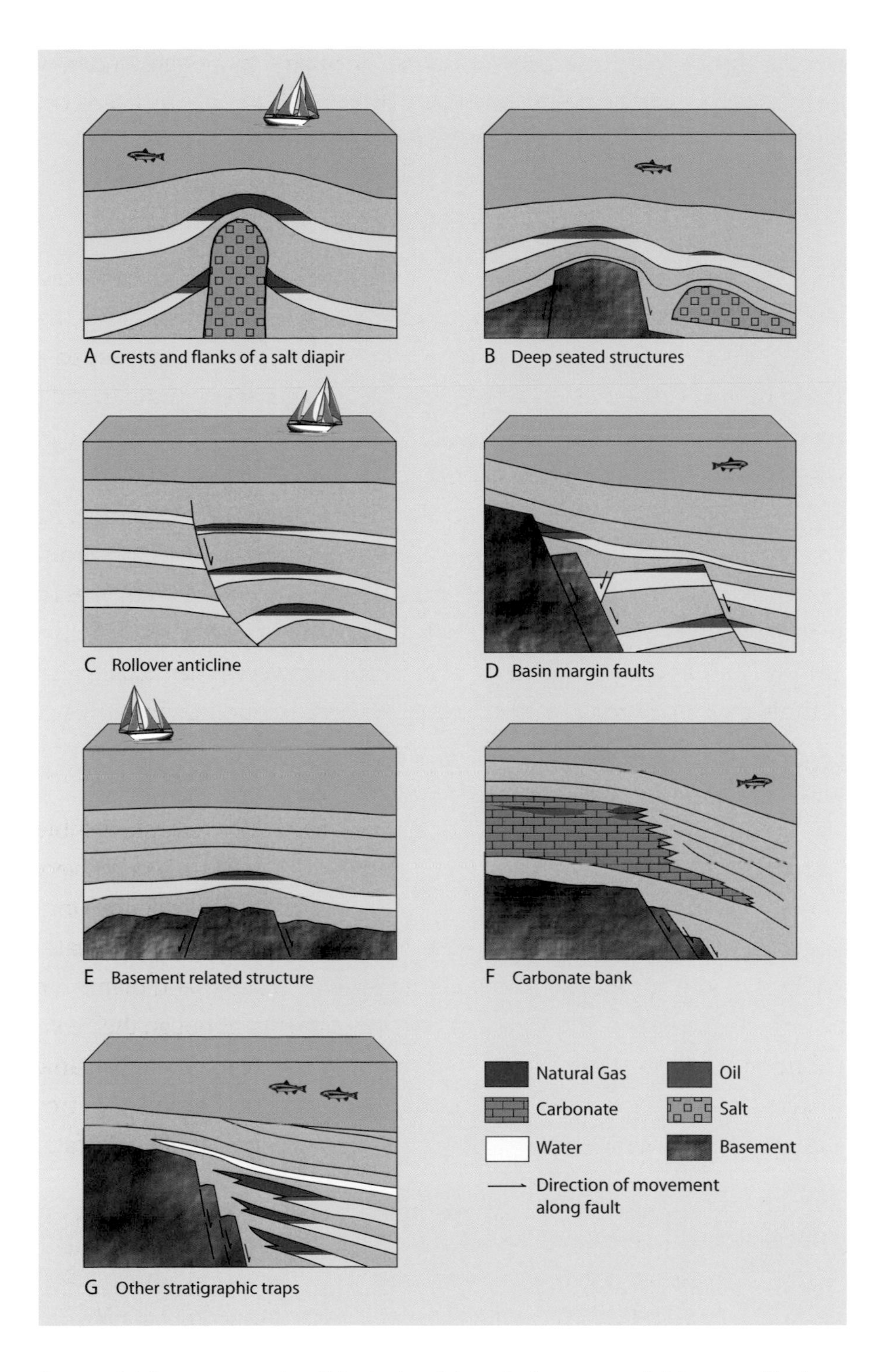

How crude oil and natural gas accumulate in traps. Oil and gas migrate upwards until they are stopped by impervious beds. Folding and faulting produce "structural traps" (A-E) and lateral changes in rock type produce "stratigraphic traps" (F-G). Anticlines are classic structural traps produced by folding of rocks. Folding is caused mainly by lateral compression (as discussed in Chapter 1). However, anticlines can also result from upthrust of underlying basement rocks (B), or of salt structures such as diapirs (A) and pillows (B). (In oil exploration, the term "basement" is used for deeper, usually harder rocks that have low potential for oil and gas.) An anticline can also be formed through draping of sediments over a pre-existing basement high (E). A "rollover" anticline (C) forms where dipping beds are faulted in a tensional (as opposed to a compressional) situation; beds on the "downthrow" side of the fault collapse into the fault zone, their dip increasing as faulting continues, so that older beds tend to be "rolled over" more strongly than younger strata. Faulted blocks can also form ideal traps for migrating oil and gas (D). Limestone that originally formed a carbonate bank on the continental shelf, like the modern Bahamas Bank, can produce excellent stratigraphic traps (F), as can sandstones enclosed by impervious clay beds (G). In all of the examples illustrated, vertical scale is greatly exaggerated.

from which most crude oil is derived. Land plant material is more likely to yield natural gas than oil.

Rocks in which petroleum forms are called "source rocks". Marine sedimentary strata—such as shales, mudstones and siltstones—are good source rocks. Oil and natural gas are found in minute spaces (pores) between rock grains. The amount of pore space in a rock is referred to as

Oil rigs in Halifax Harbour, NS.

"porosity", and the extent to which the pores are connected to each other is the rock's "permeability". This permeability controls the movement (or "migration") of the oil and natural gas through the rock.

Oil and natural gas are lighter than water and so tend to migrate upward from the source rock. Eventually, they either escape at the surface, where an oil seep forms, or they are trapped in a "reservoir rock". A "trap" or "seal" is an impermeable layer that overlies the permeable reservoir bed; it stops the oil and natural gas from escaping. In the reservoir rock, oil floats on any water present, and natural gas floats on the oil. But all the oil, gas and water is contained within the pores of the rock.

To be effective, a trap must have a particular shape. Some traps are caused by bending or folding of the strata. If such a fold is dome-shaped, like an inverted saucer, there will be an area at its crest which will hold, or "trap", migrating oil and natural gas. The hydrocarbons will rise into the crest of the structure, displacing the water. This so-called "anticlinal theory" of oil entrapment was conceived in 1842 by William Logan, the first Director of the Geological Survey of Canada. Anticlinal traps contain most of the known conventional reserves of petroleum.

In contrast to geological resources such as coal, metals, or building stone (which have been used by people for millennia), petroleum is a relative newcomer. Indeed, there was limited use for it until about 1900. The Maritimes is one of the oldest oil- and natural gas-producing regions in the world. Oil was discovered in a shallow well about 15 kilometres southeast of the city of Moncton, NB, in 1859. Fifty years later, in 1909, Maritime Oilfields Ltd. discovered the Stoney Creek oil and natural gas field south of Moncton. The reservoir consists of sandstones that were deposited in a delta along the margin of a Carboniferous lake about 360 million years ago. It produced nearly one million barrels (about 160,000 cubic metres) of oil and 30 billion cubic feet (about 860 million cubic metres) of natural gas until its closure in 1991. The natural gas was piped to the nearby town of Hillsborough, NB, as well as to Moncton, where it supplied much of the city's energy needs for nearly 50 years.

The first offshore well drilled in Canada was Hillsborough No.1, off Prince Edward Island (not to be confused with Hillsborough, NB). This

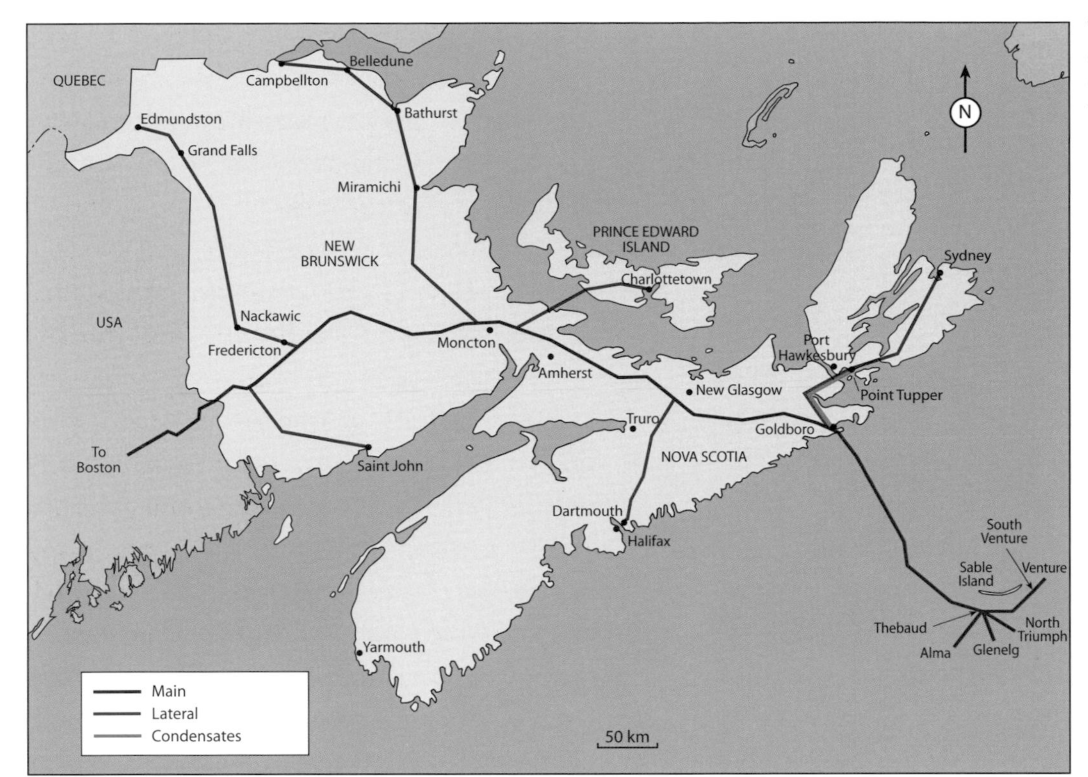

The location of gas fields offshore and of pipelines offshore and onshore in the Maritimes.

well, drilled in 1943 on an artificial island, reached a depth of about 4,500 metres and remains one of the deeper offshore wells in eastern Canada.

On the Scotian Shelf, drilling for petroleum started in the late 1960s. The big breakthrough was the discovery of the Venture Field, near Sable Island, in 1979. This, combined with five other gas fields, all in the vicinity of Sable Island, encouraged the construction of a pipeline from the production site across the Scotian Shelf, Nova Scotia and New Brunswick and into New England. Gas started flowing from these offshore fields on 31st December 1999, an auspicious day for Nova Scotia. And, from 1992 to 1999, more than 50 million barrels of crude oil, known as "Scotian Gold", flowed from the Cohasset-Panuke Field, about 40 kilometres southwest of Sable Island. This oil was transported by tanker to refineries along the eastern seaboard of North America.

The Vital Fluid

"Ground water" is water that occurs in pores and fractures in rocks, and is one of our most precious commodities. As for petroleum, factors such as porosity, permeability, migration and entrapment are important in understanding ground-water distribution. How deep can ground water be found? Amazingly, the present record in 12 kilometres, but there is evi-

Fresh water, a priceless resource.

dence that it may go even deeper. Below a depth of about 20 kilometres, however, the pressure created by the weight of the overlying rock is so great that there are unlikely to be any pores or fractures to hold fluids.

The ground-water zone is that part of the crust in which spaces in the rock are filled, or saturated, with water. The top of the ground-water zone is called the "water table". The surface of a pond or lake represents the water table at that location. Although in dry climates the water table may be a kilometre or more deep, in most places in the Maritimes it lies within a few metres of the surface. The shape of the water table generally mirrors the surface topography, but is smoother. Where there are soluble rocks, such as limestone or gypsum, ground water can occasionally collect in large underground rivers or pools. For example, in Maitland Caves, near Truro, NS, underground rivers and pools have filled caverns hollowed out of limestone.

Ground water is an important part of the hydrologic cycle, a complicated system of endlessly circulating water that also involves oceans, lakes and rivers, the atmosphere and the biosphere. If we exclude the oceans, polar ice caps and glaciers, ground water accounts for about 95 percent of the Earth's water. Ground water reservoirs, known as "aquifers", are the most prolific source of fresh water. ("Fresh water" is defined as having 100 to 500 milligrams of total dissolved solids, including salts, per litre, whereas "sea water" contains 35,000 milligrams of salt per litre.) The purity of ground water varies, depending on the minerals and elements dissolved in it as it percolates through rock. For example, ground water passing through limestone is changed to "hard" water by dissolving calcium and magnesium ions from the rock. Soap will not easily form a lather in such hard water. Pure water, as found in most rain, is "soft", because it does not contain calcium or magnesium ions and produces a beautiful lather with soap.

Because of its ability to dissolve minerals, most deeper ground water is "briny", having a salt content that can be as high as, or even higher than, that of sea water, and thus is unsuitable for drinking. Ground water also readily dissolves minerals other than salt from the rocks, and this too can lead to problems. Thus, the water from some wells drilled in the Meguma Terrane of Nova Scotia is not fit to drink, even though by definition it is "fresh", because it contains high quantities of arsenic, lead or cadmium.

Water that originated as rainfall and is stored in an appropriate aquifer provides one of the best sources of drinking water in areas where there are no large, pollution-free surface reservoirs or lakes. Prince Edward Island, for example, obtains almost all of its drinking water from aquifers in Carboniferous to Permian sandstones.

Once used or polluted, ground water from aquifers can take up to tens of thousands of years to be replenished or purified. This is because the underground water must follow a slow tortuous path through tiny cavities and cracks in the rocks. Like smoke carried downwind, ground water will disperse downstream. An aquifer must therefore be carefully managed, because if it becomes contaminated from a poorly designed landfill or hazardous waste site, it is sometimes impossible to remedy the problem. In coastal areas, ground water can often be polluted by sea water. Because sea water is denser, it will form a wedge under fresh water within an aquifer and, if not disrupted, the two systems will remain in this balanced state. But when people draw excessive amounts of fresh water from the aquifer, sea water moves upward and contaminates the wells. Such an invasion is extremely difficult to reverse.

Sometimes ground-water seepage can adversely affect the petroleum in reservoir rocks. When fresh water near the surface seeps into deep, oil-bearing reservoirs, it may bring with it a significant volume of dissolved oxygen. The oxygenated ground water lets bacteria within the Earth's crust feed on organic material, reducing (or "biodegrading") the oil to a thick, gooey tar that is expensive to recover and refine.

Sustaining Resources and Protecting Our Environment

A popular catch phrase today is the term "sustainable development". According to the 1987 Brundtland Report to the United Nations, sustainable development is development that does not compromise the ability of future generations to meet their own needs. This is extremely important when considering natural resources, such as fisheries, forests, and rocks and minerals. The sustainable development of potentially renewable resources such as fisheries and forests presents a different challenge from that posed by the use of geological resources such as coal, petroleum

The natural beauty of the Maritimes is worth sustaining.

and gypsum. These non-renewable resources, once used, cannot be rejuvenated or replenished, however well we manage them.

How does depletion of our natural resources affect us in the Maritime Provinces, where much of our economic wealth was built on coal? When coal replaced wood as a source of fuel and energy, it was easily accessible and cheap to mine in such areas as the Minto, Springhill and Sydney coalfields. The ready availability of coal also influenced other developments such as the steel industry. With prolonged use, the more mineable and thicker coal seams were exhausted, and mining became more dangerous and less economical. Today, coal in our region is giving way to oil and natural gas, which is becoming the accepted fuel for heating and electric power generation. This development will only work as long as our oil and natural gas wells don't dry up.

Another important concern when extracting rocks and minerals is that we do not permanently destroy landscapes and habitats in the process. Open pit mining is economical, but only if the land is successfully reclaimed afterward. An example of successful reclamation is in the Pictou area, where the mined landscape has been enhanced and the vegetation re-established. One problem that needs to be resolved urgently is how to avoid nightmares such as the Sydney tar ponds. And how do we reclaim contaminated areas and purify polluted ground water?

Whatever the solution to these vitally important problems, geological input is essential. By learning about how the Earth was formed, how various processes have shaped it, how climates have changed through time, and how life has evolved, geology contributes invaluably to dialogue on significant issues. Whether the topic is global warming, water resources, nuclear power generation, rising sea levels, earthquakes, massive flooding, widespread forest fires or population growth, scientific exploration of the Earth helps to supply the context and data necessary to develop appropriate social policies and to design effective technological solutions.

Soil

Soil, literally the dirt beneath our feet, is formed initially from the weathering of bedrock at the surface to produce a layer of loose material, or "regolith". The mineral composition of the bedrock and the resistance to erosion of the minerals determine the chemical composition of the regolith, which is generally a mix of clay, silt and sand.

Plants and animals are the critical factors in converting regolith to soil. Plants, especially grasses, are the main source of the organic matter that makes soil so fertile. Visible plant life is, however, only a small part of the story. Most of the metabolic activity in soil is generated by microscopic-sized organisms such as bacteria, algae and fungi. One kilogram of soil contains billions of these organisms, which convert dead organic matter to humus and change the nutrients so that these can be used by plants. The quality of a soil is a direct product of the biological activity of these microbes. Animals are also important, particularly those that burrow. In fact, earthworms and ants are the main agents in developing a soil in which mineral and organic matter is thoroughly mixed, and in which air and water can circulate.

The variation in soil type is due to several factors, of which biological activity is only one. These factors also include climate, the nature of the parent material (for example, whether it is bedrock or glacial) and the local topography, which affects weathering processes.

Soil scientists (pedologists) classify soils on their texture, chemical composition and layering. Though some soils are not layered, most have four layers, or "horizons". The uppermost layer is called the O horizon, and this is underlain from top to bottom by horizons A, B and C. These horizons, which are roughly parallel to the surface, differ from each other in composition and colour. Humus forms the O horizon. This horizon is usually dark grey to black. The underlying friable A horizon is the topsoil, a mix of mineral and organic material. The B horizon, where most roots are anchored, largely consists of mineral matter, often ground to microscopic-sized particles. Finally, the weathered layer of the underlying bedrock forms the C horizon. The thickness of these three horizons is highly variable, and their relative proportions are important—together with the organic and mineral content—in determining the richness of the soil.

Most soils in the Maritimes are formed on till. Because tills are derived from all kinds of rock, they have a very mixed composition. This is reflected in the texture and nutrient content of the overlying soil. Texture affects how well the soil drains and holds water, important information that engineers need before constructing roads or bridges.

Low sea cliff at Point Escuminac, NB, showing the gradation from unaltered bedrock beneath to soil above.

Well-drained podsols are ideal for blueberries. Colchester County, Nova Scotia.

Soils take time to develop, and soils in the Maritimes could only start to form after the ice cover disappeared at the end of the last glacial interval. In many cases, erosion is so active that soil has still not been able to develop and bedrock is exposed—this is the case, for example, at many places along the Atlantic coast of Nova Scotia, such as Peggys Cove. Where soil has developed in the Maritimes, it is mostly of the humus/iron podzol type. This means that the B horizon is rich in iron and/or humus and that minerals have been leached or removed from the A and O horizons. Such soil is usually well-drained and reddish-brown. However, on calcareous rocks, soils of this type are commonly grey. Blueberries grow on well-drained but poor podsols, especially those underlain by glacial outwash deposits and granites.

After humus/iron podzol type soils, the second most common soil in the Maritimes is termed "luvisol". This also has a high iron or humus content in the B horizon, but this horizon is usually more clayey. The clay allows luvisol to hold together well. Luvisols are generally good agricultural soils and are often found in larger floodplain areas such as along the Saint John River and in the Grand Lake area of New Brunswick. Luvisols also occur in plateau areas of the uplands and in parts of the Carboniferous lowlands.

"Gleyic" soil also occurs widely in the Maritimes. This grey to blue-grey, poorly drained soil does not have the usual four horizons. It occurs in the Fundy marshes, in peat bogs on Carboniferous rocks, and anywhere else where drainage is poor and low nutrients favour the growth of *Sphagnum* moss. Other organic-rich soils associated with bogs are mesisols and fibrisols.

The distribution of soil types in the Maritimes.

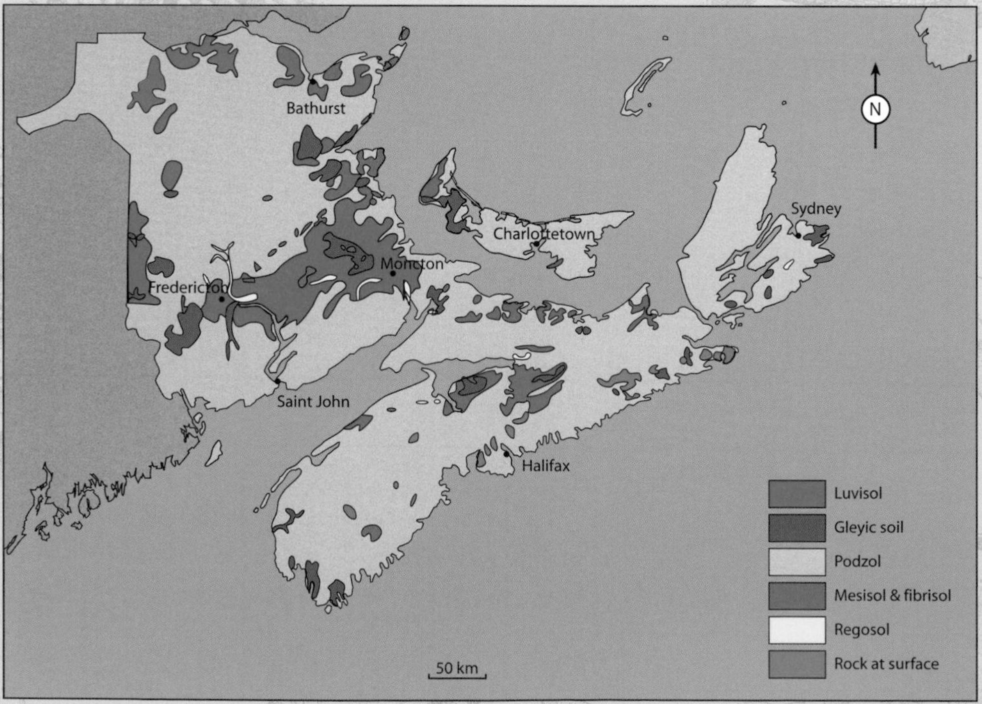

"Regosols" form by some large rivers and lakes, where there are thick deposits of unconsolidated sediment. Like gleyic soils, regosols lack true horizons. In the Maritimes, regosols occur along the Saint John and Petitcodiac rivers in New Brunswick and along the Shubenacadie and Annapolis rivers in Nova Scotia, as well as along many smaller rivers with level flood plains. Regosols also develop on active aeolian (wind) deposits, like the barrier islands off Prince Edward Island's north shore.

From Prospect to Mine— The Exploration Game

From earliest times, people have used natural materials discovered in the Earth. Pieces of rock with special characteristics (such as flint) were sought, shaped and traded for their value in making knives, spear heads and arrow points. Today, we still seek out natural resources to fashion into goods that we can use and trade. Such goods are now so commonplace that we tend to forget about the importance of minerals in our lives. Take the metallic mineral gold, for example. Aside from its beauty, the usefulness of gold was early recognized because of its malleability and indestructibility. It could therefore be fashioned into ornaments and used for currency. And it was easily mined and purified by one or two individuals with simple equipment. Hence, it was often the key to early exploration and development.

Today, the search for a mineral occurrence and the processes involved in bringing it into production are expensive and time-consuming, demanding financial backing, entrepreneurship and perseverance. Any mining venture is risky because the vast majority of prospects turn out to be too small or too low grade to be economically viable. But humans need minerals, so we have become adept at the exploration, evaluation and production game.

Mineral deposits that can be mined at a profit are called "orebodies", and rocks or minerals removed for their contained value are called "ores". An adequate grade ("assay") and size ("tonnage") is required before a mineral can be mined economically. For every producing mine, there are hundreds of mineral discoveries that do not measure up to these requirements. Indeed, the concentration of a given mineral in sufficient quantity to form an economic deposit may almost be considered a miracle. The exploration for and study of these miracles have been the focus of many geologists.

The search for a particular mineral requires a mastery of geological, geochemical and geophysical techniques. Exploration geologists often select prospective areas based on the knowledge that certain minerals usually occur in association with certain rock types; for example, tin in granites and gold in quartz veins. However, because most of the Earth's surface has already been explored, it is

Aerial view of an aggregate operation in Nova Scotia's Annapolis Valley. Sand and gravel are mined for direct use in industrial operations and for value-added uses such as cast concrete blocks and well crocks.

uncommon today to find new mineral occurrences at the surface. This means that more sophisticated techniques must be used to search underground.

The evaluation of a prospect now requires not only the detailed mapping of bedrock outcrops (geological surveys), but the systematic sampling and laboratory analysis of rock, sediment, soil, and related water samples (geochemical surveys). In addition, the work is often carried out using electronic equipment that can detect some minerals at depths of several hundred metres because they are magnetic, radioactive, relatively heavy, or electrically conductive (geophysical surveys). Aerial photography and, more recently, satellite imagery also help in defining prospects.

If any of the surveys are encouraging (for example, if the soil contains very high concentrations of copper, or if the bedrock is electrically conductive), the next step is to drill the bedrock to obtain a core sample. This is the only way of finding out the quality, extent and shape of a deposit.

In the early stages of a project, costs are relatively low, but the risk of failure is high. If initial drilling reveals high mineralization, additional exploration must be carried out to determine the limits of the mineralized zone and to calculate the tonnage and grade of mineralized rock. If tonnage is great, grade need not be high, and vice versa. Operations may cease any time that assays suggest that an economically viable deposit is unlikely to be found. Hundreds of holes may be drilled before a decision is made to mine the ore. Although with time the project becomes more expensive, the risk of failure tends to diminish. If all goes well, the mining company can expect a profit, and it is the payoff that makes the investment in dozens of failed prospects worthwhile. The time from discovery to production is never less than several years, and may be a matter of decades.

Geological and engineering factors are not the only ones that may affect the value of a mineral deposit. For each new mining proposal and operation in Canada, there must be an Environmental Impact Assessment and a reclamation plan for the time when the deposit is "mined out". To combat the poor public image of the industry resulting from past mistakes, mining companies now take their environmental responsibilities more seriously.

A mastodon wades through a pond formed in a sink hole. The vegetation consists of a sparse spruce forest, like that covering much of the Maritimes just before the last (Wisconsinan) ice advance.

CHAPTER 9
The Ice Age and Beyond

The Great Freeze and its Fluctuations

For most of Earth's history, there has probably been year-round ice and snow in highlands close to the poles. At certain times, however, ice has extended onto lowlands and lower latitudes. These episodes, when ice was more widespread and weather cooler, were the ice ages.

The most severe ice ages were probably those of the late Precambrian, about 700 to 550 million years ago. An ice age around the time of the Ordovician-Silurian boundary (about 445 million years ago), left iceberg-related "dropstones" in the Meguma Terrane rocks of Cape St. Mary, NS. And ice ages in the Carboniferous Period (about 360 to 290 million years ago) were probably the driving force for the Coal Age cyclothems.

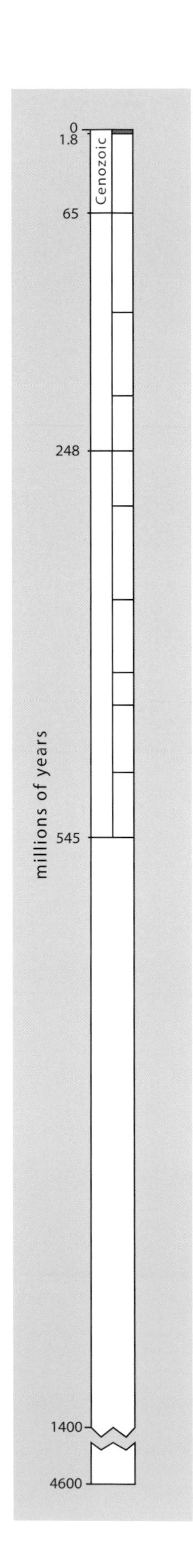

The most recent ice age, which we refer to as the "Ice Age", was the culmination of gradual cooling over the past 50 million years. By 15 million years ago, most of Antarctica was covered with ice. In the northern hemisphere, a large ice sheet first formed in Greenland, about 3 million years ago. The Ice Age is also named the "Quaternary Ice Age", since the last 1.8 million years of geological time is called the Quaternary Period. This is divided into the Pleistocene Epoch, from about 1.8 million until about 10,000 years ago, and the Holocene Epoch, from about 10,000 years ago to the present.

The idea of ice ages was born several hundred years ago, when Swiss shepherds noticed large boulders in their valley fields that had come from many miles away and could only have been moved by glaciers. (Such ice-transported boulders, known as erratics, are common in the Maritimes). However, learned men who had not seen the evidence with their own eyes were not as easily convinced. In 1830, Louis Agassiz, then a young Swiss geologist and later a renowned Harvard scholar, tried to convince his colleagues that glaciers were once more widespread than now. Convince them he finally did, but only after many years.

By the early twentieth century, the Ice Age was thought to have consisted of four glacial periods extending back about half a million years. We now know, from evidence found in deep sea cores, that there have been about 30 glacial episodes (called "glacials") within the Ice Age in the northern hemisphere, with intervening warmer spells ("interglacials"). Collectively, these glacial and interglacial episodes span the last 3 million years.

Geologists use several terms to describe ice bodies. "Glacier" denotes any large moving bodies of ice and snow. "Ice sheets" are glaciers in the form of enormous masses of ice that bury all but the highest parts of the underlying ground. "Ice caps" are small, regional ice sheets. And "valley glaciers" are glaciers confined by valley walls, commonly at the edges of ice sheets and ice caps.

During interglacials, ice covered less than one tenth of the Earth's land area, mainly in Antarctica and Greenland. But as much as a third of the landmass was covered during glacial periods, with the biggest increase in the northern hemisphere. The largest of the northern hemisphere ice sheets was the Laurentide Ice Sheet, which once covered all of North America east of the Rockies, and extended as far south as present-day St. Louis, Missouri. The four major advances of the Laurentide Ice Sheet were, from oldest to youngest, the Nebraskan, the Kansan, the Illinoian and the Wisconsinan glacials. It was the retreat of the Wisconsinan ice sheet that heralded the present (Holocene) interglacial interval. The interval since

A glacially-transported granite boulder clasped by the roots of a tree in Kejimkujik National Park, NS.

Another glacially-transported boulder from McAdam, NB, now displayed outside McAdam railway station. The unusual rock type consists of "orbicules" surrounded by granite, and was formed near the edge of a pluton—in this case, probably the Pokiok Batholith. According to current theory, veins of molten granitic material solidify, forming a narrow rim around sedimentary rock fragments. These fragments form the cores of the orbicules.

the Wisconsinan ice melted is sometimes referred to as "postglacial" time, but the evidence strongly suggests we are living at a time of temporary respite within the Ice Age, rather than at its end.

Ice ages may have been triggered by the distribution of continents and oceans, as we saw in the last chapter. But although such considerations can explain the existence of the Ice Age in a general sense, they can't explain the cycles of cold and warm periods that caused the advance and retreat

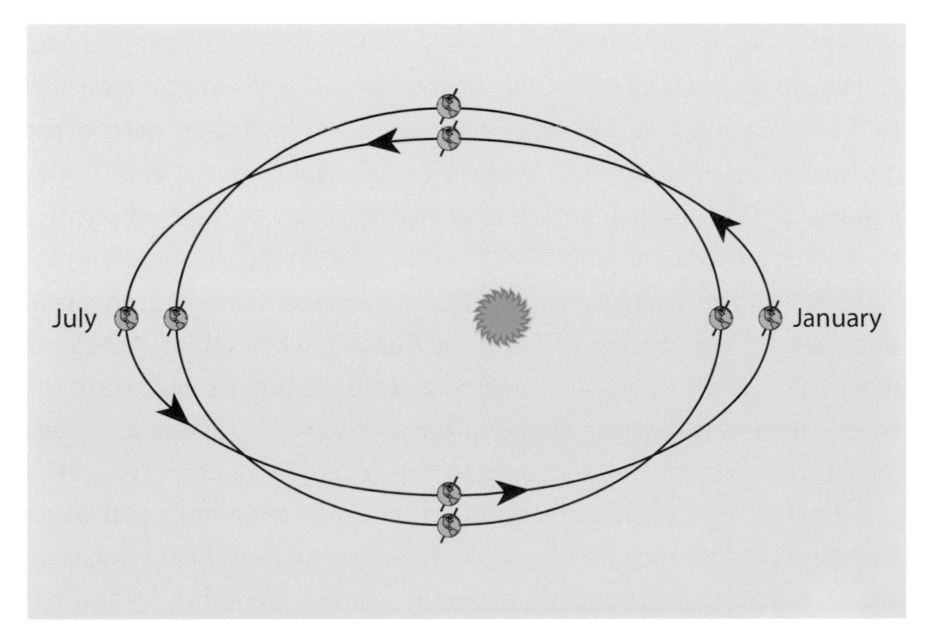

The eccentricity cycle, in which the Earth's orbit varies from an ellipse to almost a circle and back to an ellipse every 100,000 years.

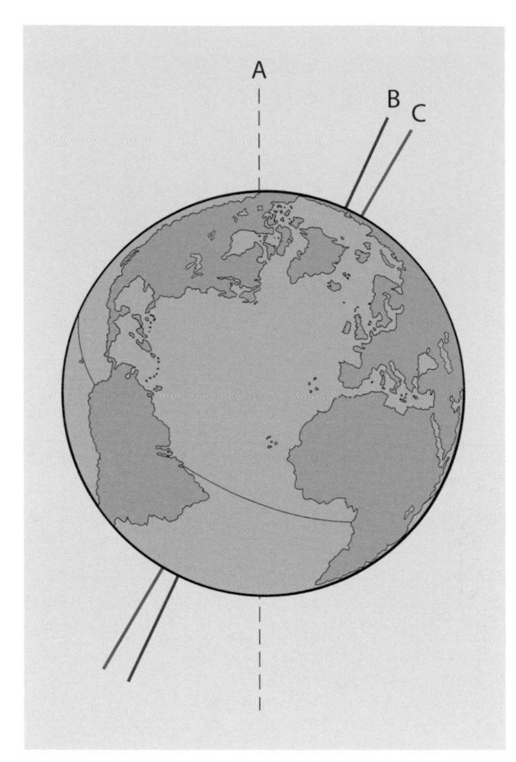

The obliquity cycle. An imaginary line (labelled A) represents the "ecliptic axis". The Earth does not rotate about the ecliptic axis, but about a "rotational axis" that is at an angle to the ecliptic axis. As shown by the two positions of the rotational axis (B and C), the angle between the rotational axis and the ecliptic axis changes over time. If the rotational axes B and C are taken to represent minimum and maximum angles of "tilt", the obliquity cycle from B to C and back to B spans 41,000 years.

of ice sheets during the Ice Age. Over the past million years, there have been climatic cycles of about 100,000 years each. During these cycles, glacials have lasted 60,000 to 90,000 years and interglacials 10,000 to 40,000 years. Why have there been these glacials and interglacials? The answer seems to be found in variations in the Earth's rotational axis and in its orbit around the Sun.

The shape of the Earth's orbit around the Sun is an ellipse, so the distance between the Earth and the Sun—and thus the energy reaching the Earth from the Sun—varies within each year. This ellipse changes from more elongate to less elongate (almost a circle) in a 95,000- to 100,000-year cycle, called the "eccentricity cycle". Thus, the annual variability in the Sun's energy reaching the Earth is itself variable over this cycle. The effect of this cycle is to intensify the seasons in one hemisphere and moderate them in the other.

In addition to this, the tilt of the Earth's axis of rotation relative to an axis perpendicular to the orbital plane (the "ecliptic axis") varies in a 41,000 year cycle, called the "obliquity cycle". It is this tilt that, at the poles, causes 24 hours of daylight in midsummer and 24 hours of darkness in midwinter. The more tilted the axis of rotation, the greater the seasonal contrast at the poles. If the axis were not tilted, uniform amounts of solar radiation would be received at each pole throughout the year, and the hours of daylight would be a constant length everywhere.

Finally, as the axis of rotation wobbles, the time of year in which any given place on Earth is closest to the sun shifts forward through the year in a 19,000- to 23,000-year cycle. This is called the "precession cycle". The two shorter cycles, the obliquity and precession cycles, are the factors controlling seasonality. At times they combine to enhance each other, whereas at other times they cancel each other out.

These cycles are known as "Milankovitch cycles", after the mathematician who first calculated them. Milankovitch cycles do not affect the total amount of solar radiation that reaches the Earth each year, but they do affect the distribution of this radiation between southern and northern hemispheres and between high and low latitudes. The amount of solar energy received in each hemisphere may vary by as much as twenty percent. In the northern hemisphere, this leads to the growth of ice sheets during intervals of cool summers and mild winters. These climate variations cause the 100,000 year climatic (or glacial-interglacial) cycles mentioned earlier.

Evidence for these climatic cycles is found in sedimentary rocks worldwide, as well as in changes in the composition of polar ice. Deep sea cores have revealed the cycles to have been very consistent over the last 800,000 years.

The Maritimes During the Pleistocene

During much of the past 2 million years, the continental shelves off Nova Scotia and the Grand Banks of Newfoundland were dry land, covered sometimes by coniferous forests and sometimes by tundra. At times, large local ice caps built up in the highlands of the Maritime Provinces, the ice being funnelled to the sea by way of deep bays and channels. Ice sheets (as opposed to these more localized glaciers) first spread over the Maritimes about 500,000 years ago. Since then, the ice sheets have probably advanced and retreated several times in the region. The locations of ice centres changed, as did the directions of ice flow. The ice was over one kilometre thick in places and flowed across the continental shelf, being halted only by the deep Atlantic Ocean. There, the ice sheet broke up into large icebergs. A mighty flow of glacier ice from Quebec gouged out the Laurentian Channel (today's Gulf of St. Lawrence and Cabot Strait), and other glaciers scoured the soft rocks on the floor of the Bay of Fundy.

Each time the ice sheets built up, they scraped and gouged the land surface, removing the soil and eroding underlying rock. And each time the ice sheets melted back, they deposited their load of soil and rock debris. During these interglacials, as in the present Holocene interval, the ice melted away, allowing the sea back in and life to return.

In the Sangamon—the last interglacial before the Wisconsinan glacial interval—the climate of the Maritimes was warmer than it is today, perhaps similar to present-day southern New England. During that time, beginning about 120,000 years ago, hardwood forests covered all three provinces. As far as we know, no humans had yet reached the Americas, although Neanderthal people were thriving in Europe and Asia.

Sea level during the Sangamon was at least 6 metres higher than it is today. In many parts of the Maritimes, this old shoreline can be seen as a wave-cut terrace in eroding cliffs, sometimes with beach sands still preserved on top. With such a high sea level, Prince Edward Island was merely a series of small islands, and an arm of the Bay of Fundy probably extended up the Saint John River valley. The Chignecto isthmus connecting Nova Scotia and

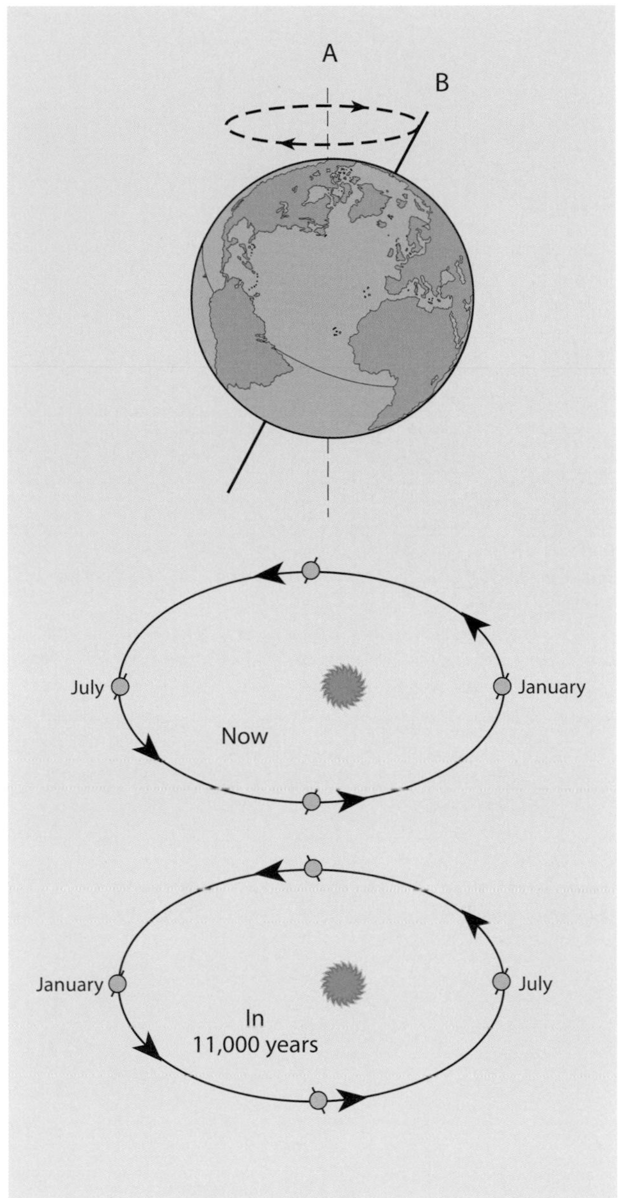

The precession cycle. The first sketch again shows the tilt of the Earth's rotational axis (B) with respect to the ecliptic axis (A). Not only does the angle of tilt vary, but the rotational axis wobbles in such a way that the relationship between the seasons and the calendar year change over a 19,000- to 23,000-year precession cycle. At present, the rotational axis is angled such that the northern hemisphere is tilted toward the Sun in July and away from it in January, as shown in the second sketch. In about 11,000 years, the rotational axis will have moved such that the northern hemisphere will be tilted toward the Sun in January and away from it in July, as shown in the third sketch. Hence, in 11,000 years, our Maritime descendants will be going to the beach in January—and at intermediate times between now and then. (For the sake of clarity, the position of the sun and the shape of the Earth's orbit are shown as unchanging in this sketch.)

The flat surface at the top of the jutting cliff is a raised beach from the Sangamon interglacial interval—the interglacial preceding the last glacial interval. The cliffs are cut in Silurian sedimentary rocks. Above the raised beach is till deposited by Wisconsinan glaciers. Arisaig, NS.

New Brunswick was also probably under water, making Nova Scotia an island.

Then, gradually, the climate cooled. Hardwood forests were replaced by softwoods and, eventually, tundra. In this colder climate, just before the onset of the Wisconsinan glaciation, mastodons roamed the woodlands and swamps, feeding on spruce trees. These elephant-like creatures sometimes became trapped in lakes or ponds, such as at Hillsborough in southern New Brunswick and at Milford in Nova Scotia, where they were fossilized and preserved, along with other animals such as turtles. Snow cover began to last through the summer in the highlands of northern New Brunswick about 70,000 years ago. Ice accumulated, and a large ice sheet eventually formed and flowed southward across the Bay of Fundy to Nova Scotia. In time, the ice became thick enough to fill the Gulf of St. Lawrence and the Bay of Fundy and flowed right across Nova Scotia and the Scotian Shelf to Sable Island and Georges Bank. This last, Wisconsinan, glacial reached its maximum extent about 24,000 years ago.

When the ice sheets began to melt back again, about 19,000 years ago, sea level started to rise. The first areas free of ice were Georges Bank, Sable Island Bank and other outer banks on the Scotian Shelf. Because land in these areas was elevated, the banks remained above sea level for thousands of years. And Georges Bank was populated by animals and plants that had spread there from ice-free areas in the United States.

Glacial Landscapes

The landscape that we see in the Maritimes today has resulted largely from the action of glaciers, including ice sheets, ice caps and valley glaciers. Ice, like water and wind, can be an agent of erosion and deposition. In the Maritimes, the overall effect of Ice Age glaciers was the erosion of soil and rock from highland areas, and its transport and eventual deposition in lowland areas.

How did ordinary snow do all this? Glaciers are moving bodies of ice and snow that form when winter accumulation exceeds summer melting ("ablation"). As snow accumulates, it becomes thicker, and its own weight caus-

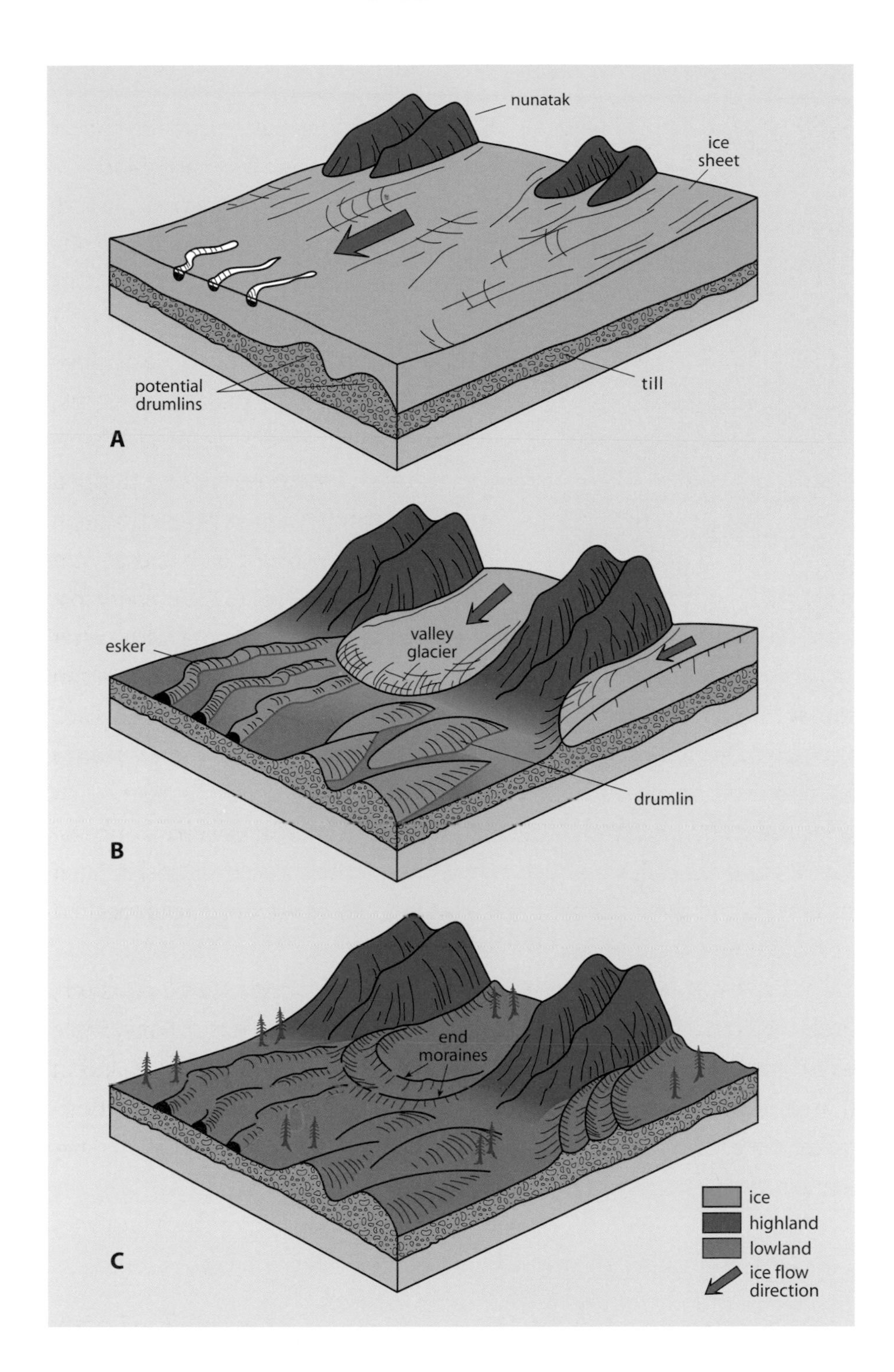

Glaciers and the landscapes they produce. A) A region covered by an ice sheet. Only the highest hills or mountains, called "nunataks", poke through the ice. The ice sheet has already deposited a thick layer of till, some of which is being molded by ice into the form of future drumlins. B) The ice has melted back, and only valley glaciers flow out from local ice centres (not shown). Eskers, formed of the sediment from streams on, in or beneath the ice, have formed ridges of well-sorted sand and gravel, and drumlins have been shaped out of the poorly-sorted till. The landscape is still mostly barren, due to the harsh climate around the glaciers, but is starting to turn green farther away from the ice. C) The glaciers have completely retreated, the land has "greened" and the valley glaciers have left end moraines at places where the melting of the glacier was delayed.

es it to recrystallize as ice. It takes decades to centuries, depending on snowfall amounts and temperature, for enough snow to build up to form ice. We know this because summer melting and winter refreezing produce layers of ice that can be used to determine age and changing climate.

On a flat surface, a thickness of about 60 metres of snow and ice will generate enough pressure for the lower part of a glacier to become plastic and start to flow. This flow generally moves at rates of a few metres a

Glacial grooves and striations in Cambrian sedimentary rocks at Hartlen Point, near Eastern Passage, NS.

Glacially polished rock face on the shoreline at Taylors Island, near Saint John, NB. The rocks result from a complex interaction of basaltic lava flows (grey) and soft muds (red), and are between 330 and 550 million years old.

A glacially-derived landscape at Peggys Cove, NS. In the foreground are glacial boulders of granite (not strictly erratics because they rest on granite bedrock, but a result of glacier action nonetheless). In the bedrock behind the boulders are examples of roches moutonnées, with smooth, gently sloping sides facing the lighthouse and steeper, more jagged slopes away from the lighthouse. The shapes of the roches moutonnées show that ice flow here was from right to left.

Three layers of till, each produced by a different ice advance, exposed in cliffs at Joggins, NS. The lowest till is brownish and represents deposits from ice flowing from New Brunswick. The middle till has a reddish tinge and represents ice flowing from Prince Edward Island. The uppermost till is the palest and represents ice flowing from the highlands of mainland Nova Scotia, to the south.

year—although when a critical mass is reached, it can be much faster. Flowing glacial ice commonly contains a lot of meltwater, which forms a film and acts as a lubricant between the ice and the underlying rock. This water repeatedly freezes and thaws as temperatures vary. As water freezes at the base of the flowing ice, soil and rock fragments stick to the glacier and are carried away. Sand and finer particles carried in the base of the ice act like sandpaper, polishing the underlying bedrock. Larger particles (such as pebbles and boulders) protrude from the base of a moving ice sheet, cutting striations, grooves and gouges in the bedrock. Landforms produced by glacial erosion usually have a smooth, humped "upstream" side and a steep, jagged "downstream" side. They are called "roches moutonnées", apparently because they resemble the backs of grazing sheep.

Sediment deposited directly from a glacier is called "till". Till usually consists of a poorly sorted mix of clay to boulder-size particles, and so was formerly called "boulder clay". But some tills contain neither clay nor boulders. Deposits of till, which cover ninety percent of the Maritimes, are called "moraines". Till that blankets bedrock is called "ground moraine", and is common throughout the three provinces.

Where glacial advance and melting cancel each other out, the edge of an ice sheet or glacier may be stationary for many years, and till can build up to form a prominent ridge. If this ridge marks the farthest extent of the glacier, it is called an "end moraine". Such a ridge or moraine system extends along much of the length of the Scotian Shelf, about 50 kilometres offshore. A ridge of till formed during the retreat of the glacier is called a "recessional moraine". (Such ridges can also form during stable intervals within a glacial advance, but the accumulation of till will be swept up as the glacier renews its forward progress.) "Ablation moraines" consist of irregular ridges and mounds formed in depressions of land by the melting of isolated, stagnant masses of glacier ice. They are particularly common in southwestern New Brunswick.

As it flows, a glacier can shape thick deposits of till into streamlined landforms, called "drumlins", which resemble the smooth back of a surfacing whale. Drumlins have played a critical role in human settlement of the Maritimes. The imposing presence of Citadel Hill, a drumlin overlooking a deep

The ridge between the two lakes is a recessional moraine in the Parrsboro valley. This ridge was deposited as the glacier retreated northward (away from the camera) about 12,500 years ago. Gilbert Lake, near Lakelands, NS.

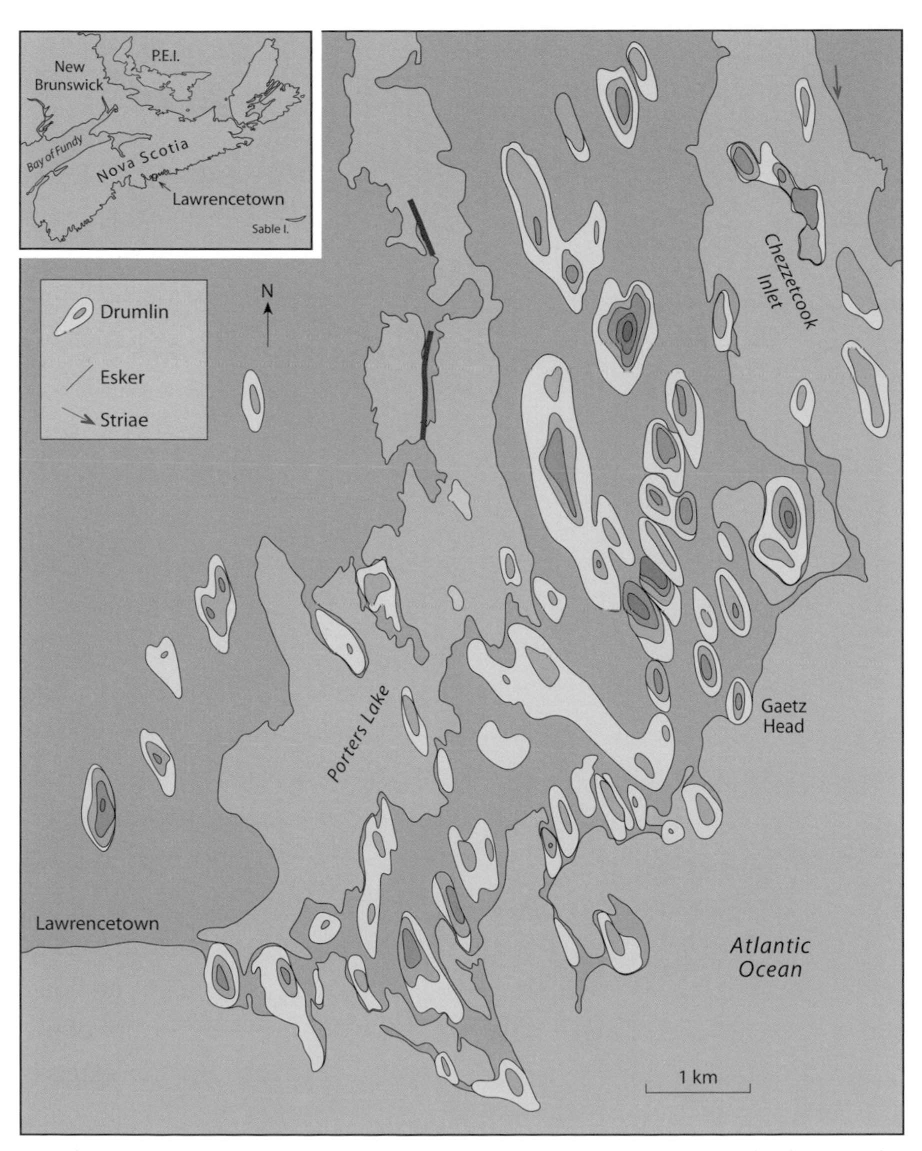

Map of the area east of Lawrencetown, NS, showing the position of drumlins and eskers. Glacial striae on the eastern side of Chezzetcook Inlet indicate the direction of ice-flow, as shown by the arrow. Notice how the shape of the drumlins generally parallels ice flow directions. Some drumlins have more complicated shapes because they were moulded by ice initially flowing in one direction, and later remolded by ice flowing in a slightly different direction.

Large, 3-kilometre long, 30-metre high drumlin, near Milford, NS, trending due south (away from the camera). This aerial view is, therefore, from the blunt, "upglacier" end of the drumlin.

An aerial view of Citadel Hill, Halifax, NS. Citadel Hill is a drumlin that was modified in the eighteenth and nineteenth centuries during the several phases of construction of the Citadel fortifications.

An esker winds its way across the landscape of Shelburne County, NS.

Glacial lake sediments, about 15,000 years old, showing annual layers ("varves"), deposited in Glacial Lake Sevogle, near Exmoor, NB.

glaciated harbour, prompted Governor Cornwallis to choose the site for Halifax, NS, in the mid eighteenth century. And drumlins provide much of the agricultural land on the otherwise unproductive, rocky, Atlantic coastal region of Nova Scotia.

Drumlins and other till-related features are deposited directly from the ice. But sediments associated with glaciers can also be transported by, and deposited from, water. Thus, flowing streams of meltwater carve tunnels in the ice, through which they transport large amounts of sand and gravel eroded from the bedrock. When the ice sheet melts, the deposits of these streams remain as ridges of sand and gravel called "eskers". Multiple eskers may form systems that extend discontinuously for hundreds of kilometres. Such a system joins the Woodstock and St. George areas in New Brunswick. Eskers and related deposits are important commercial sources of sand and gravel, as for example at Nine Mile River, NS.

Large amounts of sand and gravel carried by streams beyond the margin of the ice are called "outwash deposits". These can form extensive plains, as in the Magaguadavic River valley around Brockway, NB. Lakes are common in outwash areas, and sediments in the lake bottoms accumulate in thin, flat layers that reflect repeated seasonal run-off. Some streams build "deltas" of sand and gravel where they meet the sea or a lake. Parrsboro, NS, is built on the flat land of such a delta, and the extensive Pennfield plains east of St. George, NB, are another example.

We can determine the past extent of the ice and the patterns of its movement from several lines of evidence. Till deposits tell us that ice formerly covered an area. Pebbles of a distinctive rock type in a till, or material of a distinctive colour, show where ice has been and in what direction it flowed. Basalt pebbles from the North Mountain of Nova Scotia are found in till along much of the province's South Shore, indicating movement of ancient ice from northwest to southeast. Similarly, boulders of granite and other rocks from the Miramichi Highlands are found scattered over the Carboniferous sandstones of central and eastern New Brunswick, showing that glaciers transported them in a southerly to southeasterly direction. Erratics of granite of the type found near New Ross, NS, occur in till on top of North Mountain. This suggests that, at one time, ice must have flowed northward from the New Ross area toward the Bay of Fundy. North of Plaster Rock, NB, erratics from northern Maine indicate that some ice flowed across the Notre Dame Mountains of that state into northern New Brunswick.

Distinctively coloured material in tills can also be used to determine the history of an ice sheet. For example, near the village of Millville, NB, a

Cliffs formed from glacial outwash sediments at Five Islands, NS. Three types of sediment can be seen in the cliff. At the base is a dark-grey, flat-bedded marine clay. A middle sand unit can be identified by its diagonal cross bedding, and was deposited in a delta below sea level. The upper part of the cliff is composed of gravel deposited in streams above sea level.

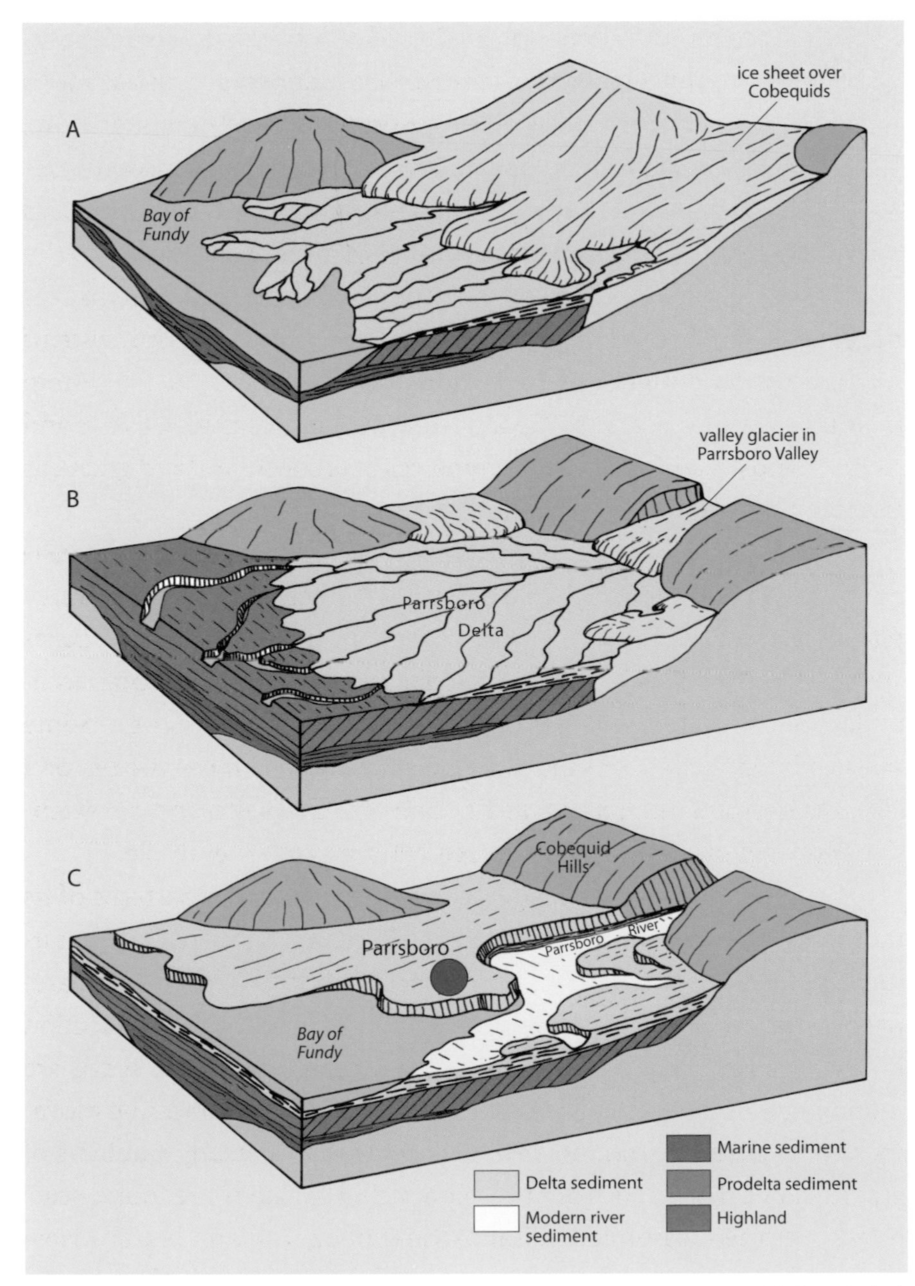

The evolution of the postglacial Parrsboro Delta. A) 15,500 years ago, the ice sheet largely covered the Cobequid Highlands, with outwash gravels deposited from streams and small deltas along a narrow coastal strip. Sea level was much higher than it is today. B) 15,000 years ago, the ice had retreated and the land had rebounded, causing sea level to drop. As a result, glacial streams cut into the delta deposits, creating the terraces visible from the highway north of Parrsboro. Valley glaciers from the Cobequids fed outwash streams, which formed a large delta (the Parrsboro Delta) in the area of the present-day town of Parrsboro. C) Today, sea level has risen, the glaciers have melted, and sediments of the delta are being eroded by tides and waves of the Bay of Fundy and the Parrsboro River.

A sand pit in the Nashwaak River Valley, north of Fredericton, NB, showing cross-bedded sands that were deposited in a delta at the margin of a glacial lake.

U-shaped "hanging" glacial valley in the distance at Bay St. Lawrence, Cape Breton Island, NS. A hanging valley is a tributary glacial valley, whose bottom is above the main valley bottom.

belt of reddish-brown till extends 27 kilometres southeastward from an area of reddish bedrock. Finally, the gouges, grooves and striations that ice left in bedrock provide evidence of the direction of ice flow. Examples are seen clearly on the granites at Peggys Cove, NS, and in the volcanic rocks at Taylors Island, Saint John, NB.

Valley glaciers cut valleys with U-shaped cross-sections, unlike the V-shaped valleys formed by rivers. Glacial valleys are perhaps not as obvious in the Maritimes as they are in more mountainous regions (such as the Rocky Mountains or western Newfoundland) but they are present. The Parrsboro and Wentworth valleys in Nova Scotia and the Saint John and Nepisiguit river valleys in New Brunswick are examples of glacial valleys.

Ups and Downs in an Ice Age World

To understand fully the evolution of the Maritimes' landscape since the last ice retreat, we must consider the rise and fall of sea level, which in turn causes coastlines to advance and retreat. There are two main processes to consider: the changing volumes of water trapped in ice sheets, and the weight of the ice sheets themselves bearing down on the crust.

As glaciers become thicker and larger, more and more water is removed from the ocean, and sea level drops. If all the ice in Greenland and Antarctica were to melt, sea level would rise by about 85 metres, drowning nearly all of Prince Edward Island and wide coastal areas of New Brunswick and Nova Scotia. In contrast, when ice sheets were most extensive, global sea level was about 120 metres lower than it is today, exposing vast areas of continental shelf. The changes in sea level caused by oceanic water being converted to ice caps, and vice versa, is thus a global phenomenon, and geologists refer to such changes as "eustatic".

In regions covered by ice, the weight of the ice sheet pushes the lithosphere down. This sinking causes material from the relatively plastic,

Changes in relative sea level over the last 15,000 years.

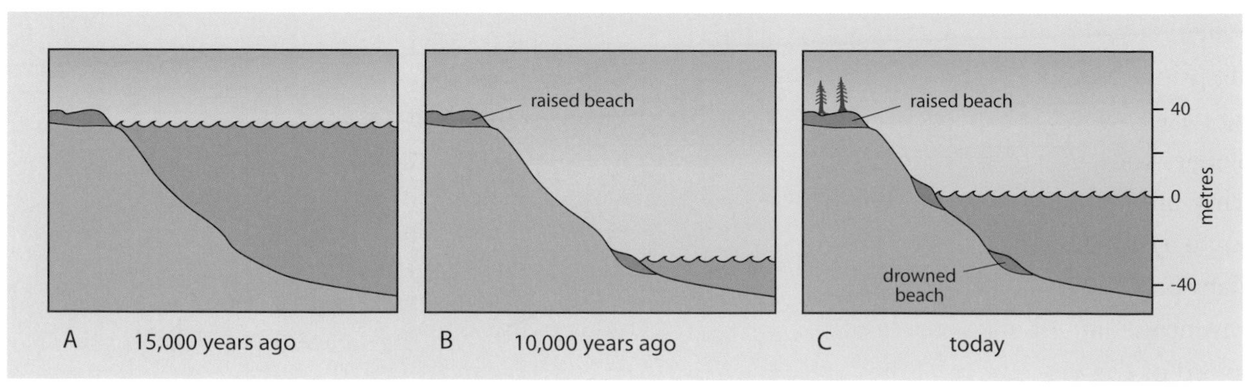

underlying asthenosphere to be squeezed outward. The land or sea floor is pushed down under the ice but rises up beyond the ice sheet. In the Maritimes, the weight of overlying ice pushed the crust down by a few hundred metres. The greatest depression was where the ice was thickest, as in central New Brunswick. There was little depression at the margins of the ice sheet, such as at Sable Island, where the ice was thin. These ups and downs of the lithosphere cause changes in sea level that are referred to as "isostatic". They can be compared to the effect of different weights on a water bed: an adult will depress the mattress much more than a small child. And the mattress will bounce back once its load is removed. Similarly, the Earth's crust bounces back, or rebounds isostatically—although very slowly—after an ice sheet melts.

The Ice Melts

The transition from barren Wisconsinan glacial ice to today's forests was marked by several abrupt climatic events. But how do we know about such climatic changes if there was no one around to record them? The answer lies in core samples of sediments laid down in lakes, bogs and nearshore oceanic environments since the ice sheets melted. As with more ancient sedimentary rocks, the oldest information is in the deepest part of the core and the most recent information is at the top. The cores almost always contain thousands of microfossils, especially pollen. The study of such fossil pollen (called "palynology") is one of the primary methods of reconstructing past environments. As those with allergies know, pollen grains from flowering plants and conifers are readily borne by wind and can settle anywhere. Many pollen grains inevitably find their way into sediment and thus provide a record of the vegetation of the time.

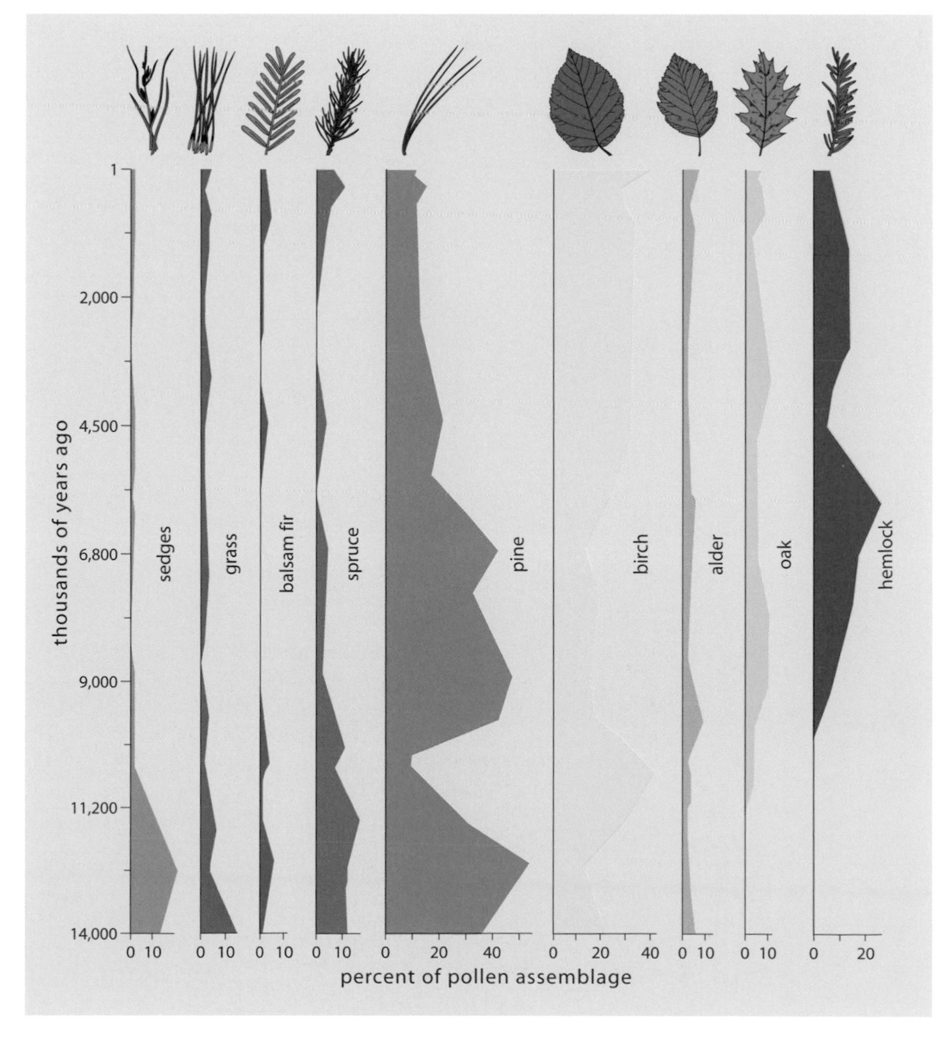

Climatic changes over the last 14,000 years are reflected in pollen types and abundances in a core from Penhorn Lake, Dartmouth, NS. Here, 14,000 years ago, there was a tundra vegetation similar to that now found in northern Quebec. The appearance of oak 11,200 years ago reflects warming and a northward migration of the boreal forest. From 9,600 years ago to the present day, the more temperate Acadian Forest has dominated, with common pine, spruce, hemlock, fir, birch and oak.

Stages in the retreat of Wisconsinan glaciers from the Maritimes.

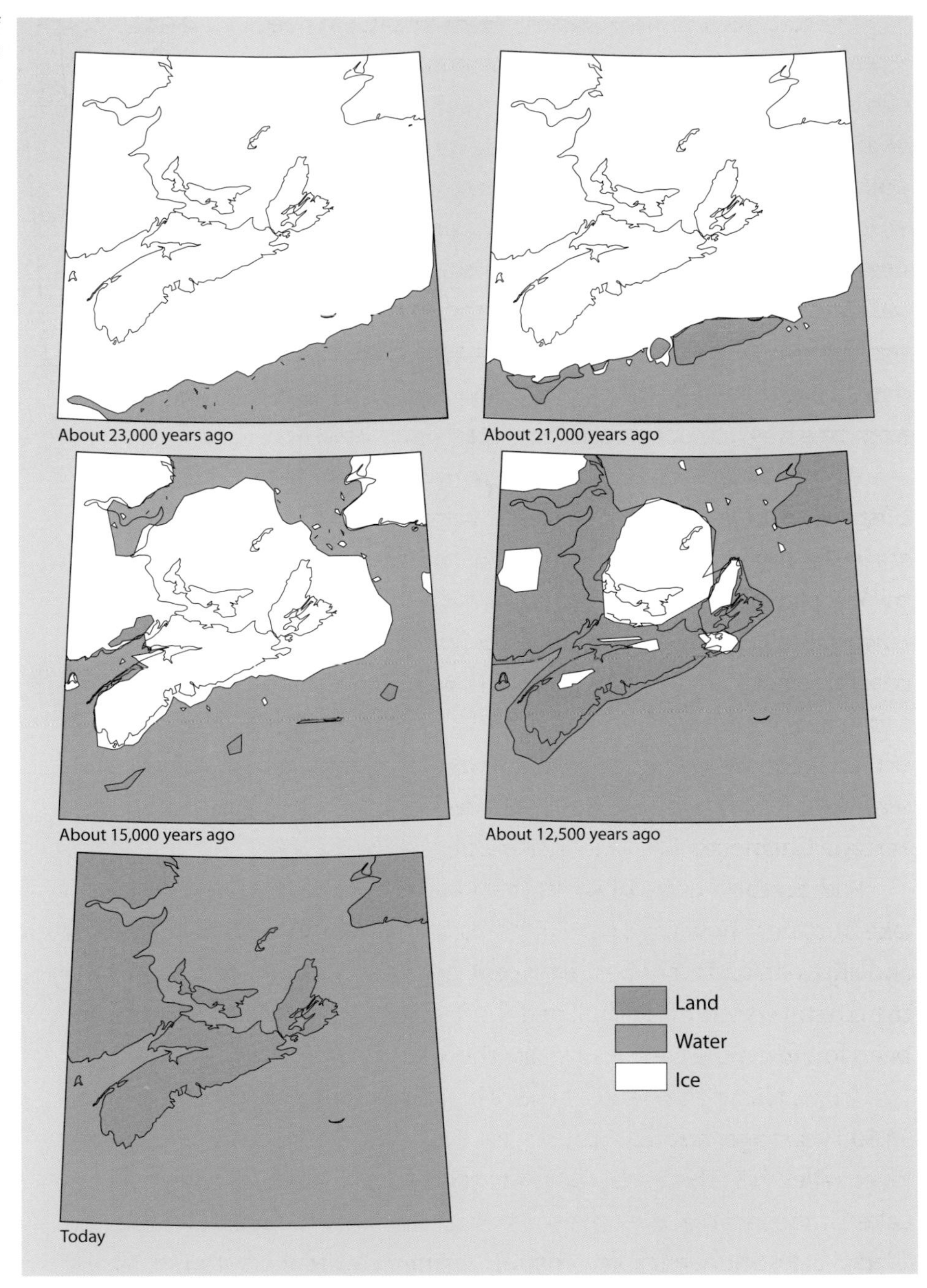

Because we know that different climatic regions have different vegetation, we can deduce past climates from the fossil pollen record.

Changes in temperature and precipitation during the growth or decline of ice sheets trigger changes in vegetation. Thus, fossil pollen grains provide information about the climate. Other fossil groups, such as insects on land and foraminifers in the ocean, also provide clues about the environment, but pollen grains are particularly valuable because they can be found in both terrestrial and marine sediments.

To establish a climatic history from fossils, we need to know the age of the sediments that contain them. Certain events within historical time, such as the introduction of alien plants by European settlers, leave a distinctive pollen record and so provide benchmark dates. However, for sediments deposited before historical records, we need other techniques. The most important of these is radiocarbon dating, which can be used to date organic materials as far back as 50,000 years ago. Radiocarbon dates have to be adjusted slightly, based on information from tree rings, to obtain the "calendar" dates used here.

Raised beach, Squally Point, near Apple River, NS. The cliffs are formed of Devonian and Carboniferous volcanic rocks, which are overlain by Holocene sand and gravel beach deposits.

About 16,000 years ago, ice retreated rapidly from offshore regions because of climatic warming and rising sea levels. But glaciers still covered most of the Maritimes, and ice was still thick in places like the Antigonish and Miramichi highlands and over Prince Edward Island. As ice melted, the sea flooded the isostatically depressed coastal regions and, locally, coastlines were very different from those of today. Beaches formed on land that was just beginning to rebound. Some of these beaches, known as "raised beaches", are now several metres above sea level—for example, at Baie du Vin Peninsula, NB. And one raised beach, at Cape Chignecto, NS, is now 38 metres above sea level.

Radiocarbon dates of shells from raised beaches and of wood from lake bottoms show that, by 15,000 years ago, the southern Bay of Fundy and adjacent land areas were ice free. Marine mammals began to populate the coastal waters of the Bay, and fossil walrus, narwhal and beluga have been found on the New Brunswick shore.

Truro, NS, at the head of the Minas Basin, became ice free around 14,500 years ago, but ice must still have blocked the Lower Shubenacadie River valley, NS. There, clay deposits testify to a great lake, called Glacial Lake Shubenacadie, a remnant of which survives today as Grand Lake. Glacial Lake Shubenacadie probably formed sometime between 14,500 and 14,000 years ago and lasted for a few hundred years. It had an outlet to the south through lakes William, Charles and MicMac between Fall River and Dartmouth, later the route of the Shubenacadie Canal. Once the ice dam was removed, the drainage returned to its former northerly direction. Clay deposits are found in the middle and lower Saint John River valley, although these sediments may, at least in part, be estuary deposits

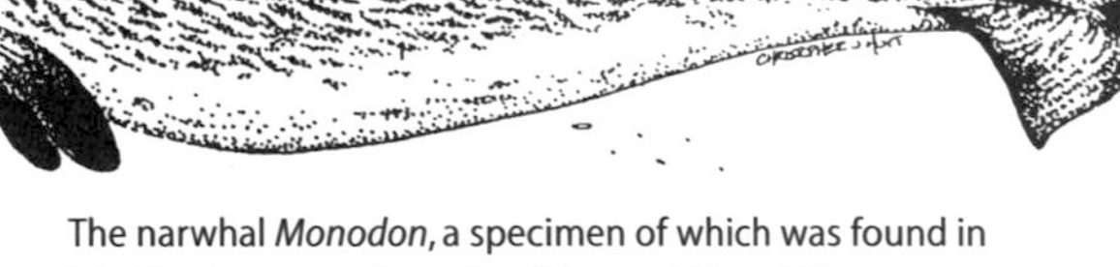

The narwhal *Monodon*, a specimen of which was found in late Quaternary sediments off Jacquet River, NB.

rather than lake deposits. Other glacial lakes formed in lowland regions throughout the Maritimes, as ice dams changed drainage routes.

At first, the postglacial landscape was like the tundra of present-day northern Quebec and Labrador. This is shown by the presence of fossil beetle species in Nova Scotia that today live around Ungava Bay in Labrador. Low shrubs and herbaceous vegetation grew where there was moisture, suitable soil and some protection. Trees began to appear after about 13,000 years ago, first in southern New Brunswick and mainland Nova Scotia, and formed open woodland of spruce, sometimes preceded by aspen. There is also evidence for caribou and other large mammals in the Bay of Fundy area. Sea level was near its lowest postglacial point—30 to 70 metres below its present level—and remains of the great Pleistocene land mammals from Georges Bank, including mastodon and mammoth, have been dated to 12,500 years ago.

This shallow section, near Lismore, NS, records rapid climate change at the end of the last glaciation. At the base are reddish glacial-lake sediments, deposited about 14,000 years ago from a glacial lake centred over the present-day Northumberland Strait. Above these sediments is a soil layer—black followed by reddish to yellow material, containing peat and wood. This soil layer represents a time, after the lake drained, of warmer climate and forested landscape, about 14,000 to 12,500 years ago. Above this soil layer is another suite of reddish lake deposits, representing a sudden cooling interval about 12,500 to 11,000 years ago; this interval is known as the "Younger Dryas". With the cooling, ice re-advanced, blocking drainage and re-establishing glacial lakes. Above these upper lake deposits is the modern soil, established over the last 11,000 years.

The first people in Nova Scotia were hunters who pursued caribou up the Bay of Fundy and who camped at Debert, NS, where their artifacts have been found. These hunters were part of the wave of immigration by the Clovis people, identified by their distinctive spear points, who spread throughout North America from Asia. They arrived in Nova Scotia around 12,500 years ago and remained for several hundred years. Paleo-Indian tools have also been found north of Fredericton and along the Fundy coast near St. Martins, NB.

By 12,500 years ago only small, isolated glaciers remained in the Maritimes. The Saint John River valley - Bay of Fundy area had been free of ice for several hundred years, and radiocarbon dates from sites in eastern Cape Breton Island suggest that the ice there had also disappeared. But then, an abrupt cooling began—the Younger Dryas event—which lasted for several hundred years. Glaciers and glacial lakes returned, the latter flooding lowland boggy areas. The clays deposited from these lakes, as well as sediments in coastal areas that draped older peat and wood beds, form a distinctive layer. Around this time there was a major extinction event across North America, with the mastodon, the woolly mammoth, the sabre-toothed cat and the bear-sized giant beaver all disappearing. Spruce trees were mostly or completely replaced by shrub tundra. The site at Debert was abandoned, and there are no other human records in the Maritimes for several thousand years—a period known to archaeologists as the "Great Hiatus".

The Younger Dryas event is a paradox because it occurred at a time when the Milankovitch cycles indicate peak warming. If scientists can explain this seeming contradiction, it may help us to better understand what happens during periods of global warming.

Into the Modern World

The woodland caribou *Rangifer*. Antlers from this type of animal were found in probable Holocene sediments off Grand Manan Island, NB.

About 11,000 years ago, after the Younger Dryas event, there was a rapid warming trend, and glaciers disappeared from the Maritimes. Forests of spruce and birch appeared, then were replaced by pine-dominated forests. Pine probably survived the frequent forest fires better than spruce: these fires left a distinctive fingerprint in the form of charcoal. During the "Hypsithermal" interval (the warmest since the ice retreated), from about 9,000 to 4,500 years ago, the temperature warmed quickly to a post-glacial maximum average of up to 2°C warmer than today. Maple and beech trees became common over much of the Maritimes, and hemlock replaced pine. Cooler and moister conditions prevailed after the Hypsithermal, especially in the last 2,000 years.

The balance of eustatic and isostatic changes was such that, about 16,000 years ago, the sea level off Halifax was 40 metres lower than it is today. By 11,500 years ago it had dropped even lower, to 65 metres below the present level, before starting to rise again. Bedford Basin, at the landward end of Halifax Harbour, was a lake until 9,000 years ago. And at Saint John, NB, a true waterfall existed at the site of the present-day Reversing Falls. Sea levels have risen again during the last few thousand years, drowning former river valleys, creating magnificent harbours, and destroying many early Mi'kmaq settlements in the process. Prince Edward Island was cut off from the mainland about 7,000 years ago. Today, sea level continues to rise along the coastline of most of Nova Scotia and southern New Brunswick. Thus, an eighteenth-century corduroy road across the Tantramar Marshes, near Aulac, NB, is now buried by salt marsh, and mooring rings in the old harbour of Louisbourg, NS, are now below high-tide level. In the Gulf of St. Lawrence, sea level continues to rise by about 2 millimetres per year. This has contributed to the rapid erosion of the coastline of Prince Edward Island—on average half a metre annually.

Continental drift and the supercontinent cycles cause climate variability on scales of tens to hundreds of millions of years. Milankovitch cycles cause fluctuations on scales of tens to hundreds of thousands of years. Historical records give accounts of climatic events which may be the latest manifestations of fluctuations on a scale of hundreds to thousands of years. The so-called "Little Ice Age" began in about 1300 AD, with a return to colder conditions, and ended in the nineteenth century. Viking settlements in Newfoundland and Greenland were wiped out by the onset of the Little Ice Age. And, at least during its waning period, Bedford Basin, NS, was frozen each year, providing skating pleasure to nineteenth-century residents. End moraines in mountain regions of the world show that valley glaciers advanced during the Little Ice Age, and there is evidence of perhaps five such advances (and hence possibly five "little ice ages") since the last glacial. The cause of these short-term fluctuations is not clear, but the answer may have something to do with sunspot activity. However, like the Younger Dryas, these little ice ages are intriguing for what they might tell us about global climate change.

The Future of the Planet

How will our climate change in the future? Will the ice sheets return and again overwhelm the Maritime Provinces? Although we can recognize cyclical variations in the Earth's climate, it is difficult to predict details for any one time. This is because oceans, atmosphere, ice sheets, life, astronomical factors and the solid Earth itself all interact in a complex manner. So when will the next Ice Age come? The amount of solar energy reaching high northern latitudes will be at a low in about the year 2025, and we may see some cooling beginning at about that time. But conditions similar to those of the last great glacial episode, the Wisconsinan, are not likely to occur for at least 10,000 years. Indeed, some earlier interglacial intervals were much longer than the Sangamon, and the present interglacial may be more like these. If so, we may expect warmer temperatures and higher sea levels in the next few thousand years.

Meanwhile, other important changes to the Earth and its atmosphere continue to affect our climate today. For example, large volcanic eruptions throw dust and chemical impurities into the upper atmosphere. These reflect some of the sun's energy back into space, so that temperatures drop.

Far more important than volcanic dust, however, is the role of carbon dioxide. This is a "greenhouse gas" that allows solar radiation to reach the Earth, but hinders the radiation of heat back into space. The amount of carbon dioxide in the atmosphere varied during the Quaternary ice age. We know this from the composition of little bubbles of air trapped in the Greenland and Antarctic ice sheets thousands of years ago. We can drill and recover cores of ice and analyze the composition of air trapped through time. This gives a historical record parallel to that preserved in microfossil shells on the ocean floor, but involving the amount of carbon dioxide in the air. The amount of carbon dioxide in the atmosphere depends on volcanic activity, oceanic temperature, biological productivity, oceanic circulation, rates of erosion and the distribution of vegetation on land, all of which are influenced by variation in the amount of solar radiation reaching Earth's surface. During the last glacial, the carbon dioxide in the Earth's atmosphere amounted to about 200 parts per million. As the ice sheets retreated and the climate warmed up, the amount of carbon dioxide in the atmosphere increased to about 280 parts per million.

Human activity has caused a rapid increase in the amount of carbon dioxide (about 360 parts per million in 1999) and other gases in the atmosphere. Since the Industrial Revolution, the burning of fossil fuels has converted ancient carbon from within the Earth into carbon dioxide. Some of this carbon dioxide is recycled back into land vegetation or marine plants by photosynthesis, but much stays in the atmosphere. And natural processes are further disrupted by the rapid destruction of forests, especially the highly productive rain forests. The continuing increase in atmospheric carbon dioxide in the last 200 years appears to be causing an average worldwide warming of climate.

What would be the consequences of global warming? The Maritimes would probably become warmer and wetter, and perhaps sea ice would disappear from the Gulf of St. Lawrence. Sea level would rise, initially probably by less than a metre, but even that much could be disastrous for cities such as Halifax, Saint John and Charlottetown, as well as for low-lying areas like Prince Edward Island. The Prairies would become drier, and the boundary of agriculture's northern limits in Canada would shift farther north. Melting of the permafrost in northern Canada and Russia would release methane, also a greenhouse gas, which would further increase the warming trend. In other parts of the world, consequences could be much more severe, causing social and economic catastrophes that would change all our lives. Global warming would likely lead to larger deserts in Africa and central Asia. During the last interglacial period, sea level was up to 6 metres higher than its present level because of the melting of the Greenland and West Antarctic ice sheets. Rapid melting of either of these ice sheets would be catastrophic, since many of the world's great cities are within 5 metres of present sea level.

Geological history can thus provide us with a window into our own future. In the geological past, population explosion and rapid environmental change led to the extinction of many species. *Homo sapiens* should heed the warning. Climate and sea level change have buffeted the Earth throughout geological history and will continue to do so in the future. Human activity may accentuate, or perhaps even reverse, natural changes, and human society will have to adapt.

As we have seen in this book, the Earth beneath our feet has evolved over hundreds of millions of years through the relentless movement of crustal plates. Climate has changed over millions of years far more dramatically than those changes predicted to occur through global warming from greenhouse gases. Species have evolved, flourished and become extinct. That is probably also the fate of the human species, so our descendants may not survive to observe the plate-tectonic changes of the next hundred million years. Over that time, the Arctic Ocean will likely enlarge through spreading of the Eurasian and American plates, changing the global pattern of oceanic circulation and possibly warming the polar areas. Australia and Africa will continue to move northward, piling up mountains in Indonesia and finally closing the Mediterranean Sea. And the Atlantic Ocean will continue to widen along the Mid-Atlantic Ridge, with subduction, at present restricted to the Caribbean, developing elsewhere along the margins of the ocean, perhaps creating volcanoes and mountains in the Maritime Provinces of Canada.

> *"What next? Geology cannot answer the question; and the geologist, as he lays down his hammer and his pen, can only utter the prayer that in the future history of this old world, in whatever of new development and higher glory its Maker may have in store for it, Acadia and its sons and daughters may bear a worthy and a happy part."*
>
> (Sir J. William Dawson, *Acadian Geology*, 1855.)

Coastlines

Aerial view of coastal marsh and former railway line.
Cole Harbour, NS.

The Ever-Changing Coastline

Coastlines mark the present boundary of the ancient struggle between land and sea. They are narrow, ever-changing, and represent a very small portion of the Earth's overall surface—but a much larger portion of our affections. The general position and shape of the coastline changes over time as sea level rises and falls. Such changes, which have been dramatic in the Maritimes over the last few thousand years, continue to be the major factor in shaping our islands, bays, inlets and peninsulas. For example, at times of higher sea level, Nova Scotia was sometimes an island, rather than the peninsula that it is today. When sea level was lower than today, major islands like Prince Edward Island and Cape Breton Island were attached to the mainland. Sable Island, that far-flung piece of Nova Scotia, is the remnant of a much larger island that has been gradually submerging over the past 17,500 years. This windswept, storm-battered outpost is composed entirely of sand, and is now only 40 kilometres long and less than 1.5 kilometre wide. It has been called the graveyard of the Atlantic because of the numerous ships that have been wrecked on its treacherous shore.

The convoluted Atlantic coastline of Nova Scotia has numerous bays and inlets that represent former valleys and lowlands—many deepened by glaciers—that have been drowned by rising sea levels over the past few thousand years. Good examples are Halifax Harbour, St. Margaret's Bay, Cole Harbour and Chezzetcook Inlet, an estuary surrounded by drumlins. In the outer reaches of Chezzetcook Inlet, drumlins act as anchor points for gravel barrier beaches that protect the estuary from the Atlantic Ocean.

One of the most dramatic examples of changing sea level involves Bras d'Or Lake on Cape Breton Island, NS. The setting of wooded hillsides, tranquil pastures and fine houses such as Alexander Graham Bell's Beinn Bhreagh, near Baddeck, give the region a lakeland look. But the seaweed, oysters and jellyfish in the water are reminders that Bras d'Or Lake is part of the sea. However, this wasn't always so. Since the last glacial, Bras d'Or Lake has been joined to the Atlantic Ocean, separated, and joined again as sea level successively rose and fell. Today, salt water enters through two narrow inlets north of Sydney and the canal at St. Peters, and reaches into the heart of Cape Breton Island through a maze of long channels.

Cliffs and sea stacks at Cape Split, near Scots Bay, NS. In the background are the Minas Channel's strong tidal currents, flowing between the Bay of Fundy (to the left) and the Minas Basin (to the right).

Coastal Materials

The coastline that we see today, therefore, is an ephemeral feature, with our beaches and cliffs being the products of the last few tens, hundreds or, at most, thousands of years. And the shore is continually changing, so that future generations will see a different coastal landscape than the one that we see. How and why do coastlines change? The two main factors, other than rise and fall of sea level, are the type of material that forms the land, and the action of the sea.

Shorelines vary according to the nature of the rock or sediment that is being eroded or deposited. Perhaps the most spectacular coastlines can be seen where resistant rocks face the onslaught of the sea and form cliffs and headlands that serve as landmarks. These resistant materials include volcanic rocks along the Bay of Fundy, such as those at New Brunswick's Fundy National Park and, in Nova Scotia, at Cape Split, Five Islands and Cape d'Or. The dramatic granite headland at Cape Chignecto, near Advocate Harbour, NS, and the cliffs and sea stacks formed in late Devonian-early Carboniferous rocks at Three Sisters, near Apple River, NS, are other examples of cliffs formed from resistant rocks. It seems odd that not all hard rocks exposed at the coast form high cliffs. For example, at Peggys Cove, NS, the low, rounded scenery formed on granite is a legacy of the erosional power of glacial ice, reminding us that it is more than just the hardness of rocks that affects the shape of coastlines.

Low cliffs are common where rocks are more easily eroded, such as the Carboniferous sandstones and shales at Cap Lumière in New Brunswick and at Sydney Mines and Point Aconi in Cape Breton Island, NS. Perhaps the best-known example of this type of low cliff is at Hopewell Cape, NB, where the Hopewell Rocks are sea stacks eroded from soft Carboniferous conglomerate. Around Sydney, NS, and along the Gulf of St. Lawrence and Chaleur Bay coasts of New Brunswick, the subdued coastal cliffs are so easily undercut by wave and tidal action that slumping and collapse are common. Equally vulnerable are the red sandstone cliffs of Prince Edward Island. This is dramatically seen at Elephant Rock, north of Tignish, an example of a sea stack with an arch, where the famous "elephant " recently lost its

Sea stacks and cliffs of late Devonian-early Carboniferous rocks. Three Sisters, near Apple River

Elephant Rock, north of Tignish, PEI: a sea stack with an arch and caves (potential future arches) formed in Permian sandstones. The "elephant" has since lost its "trunk".

Aerial view of the coastline near Elephant Rock, showing low cliffs of Permian sandstone stretching into the distance.

trunk. The Maritimes coastline abounds with caves, arches and sea stacks carved by erosional forces of the sea as they wear back cliff faces.

Glacial deposits are widespread in our region and form many low coastal cliffs and headlands. Sometimes, glacial deposits cap cliffs of older rock, as at Joggins, NS, where Ice Age till forms the higher parts of the Carboniferous "fossil cliffs". Perhaps the most unusual coastal feature in the Maritimes is at Point Escuminac, NB, where cliffs of postglacial peat rise several metres above the beach.

Cave formed in soft Triassic sedimentary rocks near Wolfville, NS, with Cape Blomidon in the distance.

Waves, Currents and Tides: The Power Supply

If rocks and glacial deposits supply materials for the shoreline, it is the sea that supplies energy to erode, transport and deposit these materials. The power of the sea is reflected in tides, waves and currents. Tidal rise and fall is due to the gravitational pull of the moon. When the moon is directly overhead, there are high tides on the near and far sides of the Earth and low tides midway between. This explains why there are roughly two tides each day. The intensity of the tides varies according to the phases of the moon. The highest tides occur when the moon is new and full; lower tides occur between these phases.

The power of the sea: waves at Peggys Cove, NS.

The most spectacular tidal range in the Maritimes—and in the world—is that of the Bay of Fundy. The difference between low and high tides varies from 3 metres at the mouth of the Bay to about 16 metres at Burntcoat Head, NS. Tides are affected by the shape of the coastline, and tend to be intensified where there is a funnel effect, as in the Bay of Fundy. However, not all deep bays have such tidal extremes, and even Fundy has not always had this great range. When the sea invaded the Bay toward the end of the last glacial interval, about 16,000 years ago, the tidal range was less than two metres. Today's spectacular tidal range developed 7,000 to 4,500 years ago, and results from changes to water depth and the shape of the coastline. Today, depth and shape combine in such a way that the natural period of oscillation of the Bay of Fundy-Gulf of Maine system is about half a day, nearly matching the half-day North Atlantic tidal cycle. (You can create an analogue by sitting in a bathtub half-full of water and rocking back and forth: the water will also slosh back and forth with a natural rhythm of a few tens of seconds.) The match between the natural rhythm of the Bay and the North Atlantic tide cycle amplifies the tidal range of Fundy.

The foundations of this Second World War observation tower (now moved) at Economy, NS, have been undermined by the tides and waves of the Bay of Fundy, as the sea encroaches onto the land.

Waves are also critical in the development of coastlines. They are generated by wind on the water surface, and their intensity and direction vary

Barrier island at Conway Narrows, north of Freeland, PEI.

Barrier island, tidal channels and lagoon. Tracadie Beach, NB.

according to the weather. As a wave approaches the shore, its lower part hits the sea bed first and is slowed by friction, but its upper part is driven forward by its own inertia as a breaker. When the breaker collapses under its own weight, it sends a mass of water (called the "swash") up the beach. The returning water is called the "backwash". The swash drives pebbles and sand shoreward and the backwash carries them seaward. When the swash is stronger than the backwash, sediment is carried landward. If waves strike the shoreline at an angle, beach material is moved in the same general direction that the waves are travelling, and "longshore drift" of the sediment occurs. A narrow ribbon of sand formed by longshore drift and extending from a headland is called a "spit". Sometimes, this ribbon of sand can connect two or more headlands forming a barrier beach, such as Martinique Beach, near Musquodoboit Harbour, NS. In other cases, the ribbon of sand is unattached, forming a "barrier island", as in parts of northern Prince Edward Island and eastern New Brunswick. Where there is abundant sediment for these beaches, they can build seaward, upward, or along the shore.

During storms, the power of both swash and backwash are much greater than usual, and the rates of deposition and erosion are magnified. Major storms, such as hurricanes and northeasters, have catastrophic effects, causing rapid and dramatic changes to the coastline. The stormiest time of year is winter, so one might think that winter weather would have the greatest impact. However, along the Gulf of St. Lawrence and the Bay of Fundy coasts, winter sea ice usually prevents wave generation by winds, therefore protecting the shore. However, if sea ice breaks up and becomes mobile during storms or spring melt, it can be driven shoreward, forming large ice ramparts that can damage wharves and other coastal structures.

Beaches

Whereas cliffs are symbols of coastal erosion, beaches are indicators of coastal deposition. Beaches in the Maritimes are commonly formed of a mixture of sand, gravel and boulders derived from glacial deposits. One of the most common beach types in our region involves a sand and gravel mixture, formed when glacial till is eroded. Along the Eastern Shore of Nova Scotia, bluffs of glacial deposits are being eroded at rates of up to 10 metres each year, providing a source of sediment for beach building. In parts of Cape Breton Island, such as Aspy Bay near Dingwall, eroded glacial deposits have been reborn as sandy beaches and dunes, which protect enclosed lagoons. Unusual triangular barrier beaches, which surround brackish water lagoons, extend into the waters of Bras d'Or Lake. These beaches are formed by storm waves that move parallel to the shore, sometimes from one direction, sometimes from the other.

The beautiful, wide, sandy beaches and coastal dunes of Prince Edward Island, northern Nova Scotia, and the Northumberland Strait shore of New Brunswick are formed from the continuing erosion of the underlying late Carboniferous and Permian sandstones. The New Brunswick coast between Miscou Island and Cape Tormentine is the longest stretch of barrier coast in Canada, with barrier beaches and spits extending across shallow drowned estuaries. The barrier beaches in Kouchibouguac National Park and the 12-kilometre-long spit at Buctouche are good examples. Where barrier beaches and spits are low, sand and gravel are pushed landward by storm waves, causing the beach to migrate inland. Barrier beaches protect brackish lagoons and freshwater lakes from the impact of oceanic waves; such beaches, lagoons and lakes are common along the Atlantic coast of Nova Scotia.

On the northern coast of Prince Edward Island, sand is carried eastward by offshore currents and is deposited off East Point, where it forms a huge, underwater bank. At Basin Head, PEI, and at nearby areas, vast amounts of sand have accumulated as successive beaches, which have been "glued" together, forming ridges partly overgrown by forest. Many barrier beaches and spits also protect small fishing harbours, as at Advocate Harbour, NS.

Sandy beaches are also found at the head of the Bay of Fundy, but the most distinctive features in the Bay are the vast sandbanks and mud flats cut by tidal channels. Twice daily, these currents rework the huge volume of reddish, muddy water brought in by the tide.

Beach of variable sediment type, derived from glacial deposits. Ingonish, NS.

Sand dunes behind the beach at Greenwich, Prince Edward Island National Park.

Marshlands

In low-lying coastal areas of the Bay of Fundy and elsewhere, sediment carried by the tide can be trapped by salt-marsh grasses such as *Spartina alterniflora* and *Spartina patens*. In this way, marshes grow upward as sea level rises. When the Acadians settled the Fundy coastal lowlands, starting in the 1600s, they built dykes to cut off the salt marsh from the sea. Parts of the original Acadian dykes are still visible on the Tantramar Marshes, near Aulac, NB. In the cliff, half a metre below the present marsh surface, are the remnants of a 300-year-old corduroy road. The system of dykes is still maintained, so that a large part of the marsh is now protected from the rising sea and provides good pasture. Today, only fifteen percent of the original marsh remains. Agricultural land, roads, and houses behind the dykes lie well below the high tide level. Much of

An aerial view of tidal flats in Shepody Bay, NB, an arm of the Bay of Fundy.

the region was flooded by the Saxby Gale of 1869, and would be flooded again if a storm sweeping up the Bay of Fundy should coincide with the highest tide.

The evolution of a marsh is "written" in the sediments. On the Tantramar Marshes, for example, sediments deposited since the end of the last glacial interval are up to 30 metres thick. Within these deposits, remnants of a 4,500 year old forest, with trunks up to a metre in diameter, are visible about two kilometres north of Fort Beausejour, NB. Cedar, hemlock, pine, fir and beech were growing on glacially-derived soil when they were buried by marine sediments that accumulated over time as sea level rose (about 30 centimetres per century) and as tidal range increased. Examples of other fossil forests—these ones about 3,600 years old—are found at Castalia Beach on Grand Manan Island, NB, and at Long Point Beach on White Head Island, off Grand Manan. However, sea level rise has not been constant. The varying rate is reflected in the alternation of silt and clay layers deposited during rapid periods of sea level rise, and peat layers formed during slower episodes.

Dykes, dams and causeways influence water levels, erosion, sedimentation and peat growth, which in turn affect fish and other wildlife, as well as human activities such as farming. For example, a 1960s causeway built for Highway 101 across the Avon River at Windsor, NS, has reduced the strength of the tidal currents, so that at least 10 metres of mud has accumulated. These mud flats have now built up almost to high-tide level and are becoming covered with vegetation. Thus, a new salt marsh is forming.

The coast that we see today—its pattern of islands, inlets and headlands—represents merely a snapshot in time, a pause in the continuing changes in sea level. 8,000 years ago, the coast was many kilometres seaward. About 5,000 years ago, the Northumberland Strait was a long valley draining to the east. And, in a few thousand years, much of the coast will again have migrated inland as sea level rises. One thing we can be sure of is that the eternal battle between the land and the sea and the constantly changing coastline will always be full of surprises, beauty, and inspiration.

Backpack

Listed below are resources that provide additional information on themes covered in this book. Those items marked with an asterisk (*) are also available in French. For a more complete listing of resources, visit the EarthNet website at agc.bio.ns.ca/EarthNet.

Books and Pamphlets

— *Dawning of the Dinosaurs: The Story of Canada's Oldest Dinosaurs*, by Harry Thurston, co-published by Nimbus Publishing and The Nova Scotia Museum in 1994

— *For Love of Stone. Volume I: The Story of New Brunswick's Building Stone Industry*, by G.L. Martin, published by the New Brunswick Department of Natural Resources and Energy in 1991*

— *For Love of Stone. Volume II: An Overview of Stone Buildings in New Brunswick*, by G.L. Martin, published by the New Brunswick Department of Natural Resources and Energy in1991*

— *Fossil Hunter: Will Matthew and the Giant Trilobite*, by Randall Miller, illustrated by Judi Pennanen, published by the New Brunswick Museum in 1999*

— *Into the Dinosaurs' Graveyard. Canadian Digs and Discoveries*, by David Spalding, published by Doubleday Canada in 1999

— *The Fossil Cliffs of Joggins*, by Laing Ferguson, published by the Nova Scotia Museum in 1988

— Booklets and pamphlets of general interest are listed on the websites of the New Brunswick Department of Natural Resources (www.gnb.ca/0078/minerals/select.htm) and the Nova Scotia Department of Natural Resources (www.gov.ns.ca/natr/meb/pubs2.htm#publications).

Events, Classes and Workshops

— Fundy Geological Museum, Parrsboro, NS. Telephone (902) 254-3814. Website: museum. gov.ns.ca/fgm/programs/programs.html

— Nova Scotia Gem and Mineral Show, Parrsboro, NS. Held third week of August. Contact the Fundy Geological Museum (above) for information.

— Nova Scotia Museum of Natural History, Halifax, NS. Telephone (902) 424-7353. Website: museum.gov.ns.ca/mnh/index.htm

Background: Balancing Rock on Long Island, NS, is formed of columnar North Mountain basalt.

Field Guides

— *Discovering Rocks, Minerals and Fossils in Atlantic Canada: A Geology Field Guide to Selected Sites in Newfoundland, Nova Scotia, Prince Edward Island and New Brunswick*, by Peter Wallace, published by the Atlantic Geoscience Society, in 1998. Available from the Atlantic Geoscience Society.

— Nova Scotia Department of Natural Resources Walking Tours. Available from the Halifax office of the Nova Scotia Department of Natural Resources or online at www.gov.ns.ca/natr/meb/pubs2.htm#publications.

Museums, Displays and Centres

— Cape Breton Miners' Museum, Glace Bay, NS. Telephone (902) 849-4522. Website: highlander.cbnet.ns.ca/cbnet/comucntr/miners/museum.html

— Discovery Centre, Halifax, NS. Telephone (902) 492-4422. Website: www.discoverycentre.ns.ca

— Fundy Geological Museum, Parrsboro, NS. Telephone (902) 254-3814. Website: www.fundygeomuseum.com

— Hopewell Rocks Interpretation Centre. Telephone (toll-free) 1-877-734-3429. Website: www.hopewellrocks.com

— Joggins Fossil Centre. Telephone (902) 251-2727

— Mastodon Ridge, Stewiacke, NS. Telephone (902) 639-2345. Website: www.mastodonridge.com

— Miguasha Museum, Nouvelle, PQ. Telephone (418) 794-2475

— Moncton Museum, Moncton, NB. Telephone (506) 859-2648

— New Brunswick Mining and Mineral Interpretation Centre, Petit-Rocher, NB. Telephone (506) 542-2672

— New Brunswick Museum, Saint John, NB. Telephone (506) 643-2300. Website: www.gov.nb.ca/0130/index.html

— Nova Scotia Museum of Natural History, Halifax, NS. Telephone (902) 424-7353. Website: museum.gov.ns.ca/mnh/index.htm

— Ovens Natural Park, near Lunenburg, NS. Telephone (902) 766-4621.

— Parrsboro Rock and Mineral Shop and Museum, Parrsboro, NS. Telephone (902) 254-2981

— Science East, Fredericton, NB. Telephone (506) 457-2340

— Springhill Miners` Museum, Springhill, NS. Telephone (902) 597-3449 (summer); (902) 597-8614. Website: fox.nstn.ca/~jdemings/smm.html

A number of provincial and national parks in the Maritime Provinces have interpretive displays and/or printed materials describing the geology.

Organizations/Information Sources

— Atlantic Geoscience Society. Website: is.dal.ca/~walla/ags/ags.htm

— Geological Survey of Canada (Atlantic), Bedford Institute of Oceanography, Dartmouth, NS. Telephone (902) 426-8513. Website: agc.bio.ns.ca

— New Brunswick Department of Natural Resources and Energy. Telephone (506) 453-2206. Website: www.gnb.ca/0078/

— Nova Scotia Department of Natural Resources, Halifax, NS. Telephone (902) 424-8633. Website: www.gov.ns.ca/natr/

— New Brunswick Mining Association. Telephone 506-857-3056

— Universities and colleges in the Maritimes offering geology or earth science courses:

Acadia University, Wolfville, NS

College of Geographic Sciences, Lawrencetown, Annapolis County, NS

Dalhousie University, Halifax, NS

Saint Francis Xavier University, Antigonish, NS

Saint Mary's University, Halifax, NS

University College of Cape Breton, Sydney, NS

University of New Brunswick, Fredericton campus, NB

University of New Brunswick, Saint John campus, NB

Maps

— Geological Highway Map of New Brunswick and Prince Edward Island by Laing Ferguson and L.R. Fyffe, published by the Atlantic Geoscience Society, in1985. Available from the New Brunswick Department of Natural Resources and Energy.*

— Geological Highway Map of Nova Scotia (Second Edition) by Donohoe, H.V., Jr., and Grantham, R.G., published by the Atlantic Geoscience Society and Nimbus Publishing, in 1989. Available from Nimbus Publishing, Halifax NS or bookstores throughout Nova Scotia.

Resources for Teachers

— Atlantic Geoscience Society Video Series. See Videos below.

— EarthNet: A Virtual Resource Centre (agc.bio.ns.ca/EarthNet)*

— Fossils of Nova Scotia (museum.gov.ns.ca/fossils/index.htm)

— Fundy Geological Museum Schools Program, Parrsboro, NS. Telephone (902) 254-3814. Website: www.fundygeomuseum.com

— New Brunswick's Mineral Wealth Teacher's Handbook and CD-ROM. Telephone 506-861-9073*

— Nova Scotia EdGEO Workshop Program. Telephone (902) 426-4386. Website: agc.bio.ns.ca/schools/edgeo/edgeo.html

— Nova Scotia Museum of Natural History, Halifax, NS. Geology equipment for class field trips and activity kits can be borrowed from School Loans. Telephone (902) 424-6524. Website: museum.gov.ns.ca/mnh/educ/index.htm

— New Brunswick Museum, Saint John, NB. Tours of the geology gallery and loans of activity kits can be arranged. Telephone (506) 643-2300. Website: www.gov.nb.ca/0130/index.html

Videos

— Aerial Videos of Canada's Coasts. Website: agc.bio.ns.ca/pub-prod/coastal/coastal.html

— Atlantic Geoscience Society Video Series. Website: is.dal.ca/~walla/ags/ags.htm

All videos except Mineral Wealth of Atlantic Canada have accompanying guides. Video titles:

Offshore Oil and Natural Gas*

Mineral Wealth of Atlantic Canada*

The Appalachian Story

The Recent Ice Age*

Websites

— A Virtual Field Trip of the Landscapes of Nova Scotia (www.gov.ns.ca/natr/meb/field/start.htm)

— Bottom of Halifax Harbour (agc.bio.ns.ca/pubprod/of3154/html_bm1411/bm141101_home.html)

— Canadian Landscapes (sts.gsc.nrcan.gc.ca/clf/landscapes.asp)

— Careers in Geoscience (sciborg.uwaterloo.ca/earth/geoscience/careers.html)

— EarthNet: A Virtual Resource Centre (agc.bio.ns.ca/EarthNet)*

— Fossils of Nova Scotia (museum.gov.ns.ca/fossils/index.htm)

— Fundy Geological Museum (www.fundygeomuseum.com/)

— Hopewell Rocks: The Ocean Tidal Exploration Site (www.hopewellrocks.com)

— National Parks of Canada (parkscanada.pch.gc.ca/np/np_e.htm)

— New Brunswick Department of Natural Resources (www.gnb.ca/0078/)

— New Brunswick Museum (www.gnb.ca/0130/index.html)

— Nova Scotia Department of Natural Resources (www.gov.ns.ca/natr/)

— Nova Scotia Museum of Natural History (museum.gov.ns.ca/mnh/index.htm)

— Temperature Rising (www.climatechangecanada.org)

Credits

Paintings:

Judi O. Pennanen: p.1, p.39, p.61, p.77, p.89, p.91, p.101, p.113, p.125, all courtesy of New Brunswick Museum; cover, p.5, p.27, p.36, p.147, p.177, all courtesy of Geological Survey of Canada (Atlantic).

Stephen F. Greb: p.48.

Sketches:

Christopher Hoyt: p.42 (both), p.43 (both), p.45 (both), p.46 (all), p.49, p.50, p.51, p.52 (both), p.53, p.55, p.56, p.59, p.70, p.78 (both), p.79 (both), p.90, p.92, p.93, p.104, p.113, p.114, p.115 (both), p.120, p.122 (all), p.123 (both), p.124, p.191, p.193, all courtesy of New Brunswick Museum; p.57, p.130, p.131, p.132, p.133 (both), all courtesy of Atlantic Geoscience Society.

Photographs:

Sandra M. Barr: p.66 (bottom), p.67.

David E. Brown: p.108, p.127 (right).

John H. Calder: p.39, p.106 (top).

Howard V. Donohoe Jr.: p.11, p.13 (bottom), p.148, p.175, p.201 (top).

Robert A. Fensome: p.14 (top left, middle, bottom), p.15, p.16 (bottom), p.17, p.18 (bottom), p.30 (both), p.35 (top), p.60, p.69 (top), p.70, p.72 (both), p.75, p.80, p.82 (left), p.84 bottom), p.85 (both), p.86, p.88, p.90 (right), p.98, p.103, p.109, p.114 (top), p.116 (bottom), p.127 (left), p.128 (top), p.129 (bottom), p.131 (top), p.144, p.152 (bottom), p.157, p.159 (top), p.161 (both), p.163, p.164 (top), p.164 (bottom), p.170, p.172, p.173 (both), p.184 (top left), p.184 (top right), p.184 (middle).

Laing Ferguson: p.115 (bottom).

Leslie R. Fyffe: p.179 (right).

Ronald Garnett/Airscapes: p.14 (top right), p.16 (top left), p.66 (top), p.112 (top), p.117, p.150 (bottom), p.158, p.164 (middle), p.165, p.185 (bottom), p.198 (middle and bottom right), p.200 (top and middle), p.201 (bottom), p.202.

Patricia G. Gensel: p.50, p.93 (both).

Martin R. Gibling: p.18 (top), p.41 (bottom), p.107 (top), p.128 (bottom right).

Robert G. Grantham: p.29 (top left and top right), p.92.

Lubomir F. Jansa: p.29 (bottom).

John P. Langton: p.83.

William C. MacMillan: p.107 (bottom).

R. Andrew MacRae: p.146.

Gwen L. Martin: p.159 (bottom), p.160.

Krista L. McCuish: p.13 (top), p.99 (both), p.100 (bottom).

Steven R. McCutcheon: p.82 (right), p.154.

Ronald Merrick: p.100 (top), courtesy of Nova Scotia Museum of Natural History.

Brent V. Miller: p.64 (both).

James A. Miller: p.97 (bottom), courtesy of Alan Ruffman.

H.M. Mosdell: p.97 (top), courtesy of Alan Ruffman.

Michael A. Parkhill: p.186 (bottom).

Robert P. Raeside: p.84 (top).

Allen A. Seaman: p.162 (top), p.187 (middle).

Shaw Resources Ltd.: p.138, courtesy of Gordon Ritchie.

Ian S. Spooner: p.35 (bottom).

Ralph R. Stea: p.172 (bottom), p.182, p.184 (bottom), p.185 (top and middle), p.186 (top), p.187 (top and bottom), p.191, p.192.

Robert B. Taylor: p.87, p.197, p.199 (bottom).

Gilbert van Ryckevorsel: p.105.

Keith Vaughan: p.106 (bottom), p.112 (bottom), p.128 (bottom left), p.168, p.179 (left), p.198 (bottom left), p.199 (middle), 203.

John W.F. Waldron: p.155.

John Wm. Webb: frontispiece, p.129 (top), p.198 (top), p.199 (top).

Heinz F. Wiele: p.24 (all), p.25 (both), p.26 (all), p.40 (both), p.41 (top), p.44 (both), p.45, p.47, p.49, p.52, p.53, p.69 (bottom), p.78, p.90 (left), p.114 (bottom right and left), p.115 (top), p.116 (top), p.123 (both), p.124, p.129 (middle), p.131 (bottom), p.132, p.141, p.145, p.149, p.150 (top), p.151, p.152 (top).

Alex A. Wilson: p.156.

Schematic Diagrams:

p.4: Christopher Hoyt, courtesy of Atlantic Geoscience Society.

p.6: Walter R. Roest.

p.7: Walter R. Roest.

p.8 : Adapted from Cox and Doell (1960) (see references below).

p.10: Adapted from Colman-Sadd and Scott (1994), with input by John A. Wade.

p.12: Adapted from Colman-Sadd and Scott (1994).

p.17: John A. Wade, with input by Graham L. Williams.

p.19: Adapted from Veevers (1990).

p.20: From van Andel (1985). Reprinted with the permission of Cambridge University Press.

p.22: Based on an original in Colman-Sadd and Scott (1994), with input by Ross R. Boutilier.

p.30: Jonathan Bujak and Graham L. Williams.

p.32: Graham L. Williams, with input by Robert A. Fensome.

p.34: Graham L. Williams.

p.37: Sketches from Dawson (1868).

p.38: Sketch from Dawson (1868).

p.42: Robert A. Fensome and R. Andrew MacRae.

p.43: Robert A. Fensome and R. Andrew MacRae.

p.47: Robert A. Fensome and R. Andrew MacRae.

p.51: Robert A. Fensome and R. Andrew MacRae.

p.58: Adapted from Cowie (1974), with input by Andrew S. Henry and R. Andrew MacRae.

p.63 (top) Cees R. Van Staal and Conall Mac Niocaill, with input by Leslie R. Fyffe and Sandra M. Barr.

p.62: William C. MacMillan.

p.63 (bottom) Leslie R. Fyffe and Sandra M. Barr.

p.65: Cees R. Van Staal and Conall Mac Niocaill, with input by Leslie R. Fyffe and Sandra M. Barr.

p.66: Andrew S. Henry, with input by Leslie R. Fyffe and Sandra M. Barr.

p.74: Robert A. Fensome, with input by Leslie R. Fyffe and Sandra M. Barr.

p.76: Robert A. Fensome, with input by Leslie R. Fyffe and Sandra M. Barr.

p.78: William C. MacMillan.

p.79: Cees R. Van Staal and Conall Mac Niocaill, with input by Leslie R. Fyffe and Sandra M. Barr.

p.80: Cees R. Van Staal and Conall Mac Niocaill, with input by Leslie R. Fyffe and Sandra M. Barr.

p.81: R. Andrew MacRae.

p.82: Cees R. Van Staal and Conall Mac Niocaill, with input by Leslie R. Fyffe and Sandra M. Barr.

p.83: Cees R. Van Staal and Conall Mac Niocaill, with input by Leslie R. Fyffe and Sandra M. Barr.

p.86: Cees R. Van Staal and Conall Mac Niocaill, with input by Leslie R. Fyffe and Sandra M. Barr.

p.94 (both) Andrew S. Henry, with input by Sandra M. Barr, Howard V. Donohoe, Jr., Leslie R. Fyffe and David J.W. Piper.

p.96: Gordon N. Oakey, David J.W. Piper and Andrew S. Henry.

p.102: William C. MacMillan.

p.103: Christopher R. Scotese (Paleomap Project), with input by John A. Wade, Andrew S. Henry, William C. MacMillan and Graham L. Williams.

p.104: Martin R. Gibling, with input by Clint St. Peter.

p.105: Martin R. Gibling, with input by Clint St. Peter, based in part on an original in Howie (1985).

p.108 : Robert G. Grantham.

p.109: Christopher R. Scotese (Paleomap Project), with input by John A. Wade, Andrew S. Henry, William C. MacMillan and Graham L. Williams.

p.110: Martin R. Gibling, with input by Clint St. Peter.

p.111: Andrew S. Henry, with input from Robert A. Fensome, Martin R. Gibling, R. Andrew MacRae and Graham L. Williams, based on an original in Ferguson (1988).

p.114 (top left) R. Andrew MacRae.

p.117: Christopher R. Scotese (Paleomap Project), with input by John A. Wade, Andrew S. Henry, William C. MacMillan and Graham L. Williams.

p.118: From Leidy (1854), with permission from Academy of Natural Sciences, Philadelphia.

p.119: Andrew S. Henry, with input by Howard V. Donohoe, Jr., Leslie R. Fyffe and David J.W. Piper.

p.126: William C. MacMillan.

p.127: John A. Wade.

p.134: Christopher R. Scotese (Paleomap Project), with input by John A. Wade, Andrew S. Henry, William C. MacMillan and Graham L. Williams.

p.135 (both) John A. Wade.

p.136: Christopher R. Scotese (Paleomap Project), with input by John A. Wade, Andrew S. Henry, William C. MacMillan and Graham L. Williams.

p.137: John A. Wade.

p.140: Christopher R. Scotese (Paleomap Project), with input by John A. Wade, Andrew S. Henry, William C. MacMillan and Graham L. Williams.

p.141: John A. Wade.

p.142: Jonathan P. Bujak.

p.143: Andrew S. Henry, with input by John A. Wade.

p.153: Reginald A. Wilson.

p.165 : William C. MacMillan.

p.166: John A. Wade.

p.167: John A. Wade and G.L. Williams.

p.169: Adapted from the poster Sable Gas Project: used with permission–copyright Ocean Resources, P.O. Box 2705, Halifax, NS B3J 3P7

p.174: Antonius G. Pronk.

p.178: William C. MacMillan

p.179: Graham L. Williams.

p.180: Graham L. Williams.

p.181: Graham L. Williams.

p.183: John Shaw.

p.185: Ralph R. Stea.

p.187: Ralph R. Stea.

p.188: John Shaw.

p.189: Francine M. McCarthy with input from Robert A. Fensome and William C. MacMillan.

p.190: Ralph R. Stea.

p. 212 (both) Andrew S. Henry.

Specimens:

Academy of Natural Sciences, Philadelphia: p.118.

Robert A. Fensome: p.129 (middle).

Fundy Geological Museum: p.24 (bottom middle; Barry Iver Collection), p.24 (bottom right; Kathleen Waugh Collection), p.25 (top; Barry Iver Collection), p.26 (bottom), p.115 (bottom), p.144.

Patricia G. Gensel: p.50, p.93 (both).

Eldon George: p.108.

Peter Hacquebard: p.52.

Stephen Horne: p.44 (both).

David Mossman: p.49.

New Brunswick Museum: p.69 (bottom), p.78, p.114 (bottom), p.121, p.123 (top), p.145.

Nova Scotia Museum of Natural History: p.25 (bottom), p.40 (top and bottom), p.41 (top), p.47, p.53, p.90 (left), p.123 (bottom), p.124, p.131 (bottom), p.132, p.141, p.146, p.149, p.151, p.152 (top), p.156.

Gordon N. Oakey: p.24 (bottom left), p.26 (top and middle), p.45, p.115 (top), p.150 (top).

Redpath Museum, McGill University: p.116 (top).

Donald Reid: p.114 (bottom right).

Affiliations

Acadia University: Sandra M. Barr, Robert P. Raeside, Ian S. Spooner.

Bedford Institute of Oceanography: Heinz F. Wiele.

Brock University: Francine M. McCarthy.

Canada-Nova Scotia Offshore Petroleum Board: David E. Brown.

Columbia University: Paul E. Olsen.

Dal-Tech: Henrietta Mann.

Dalhousie University: D. Barrie Clarke, Martin R. Gibling, Krista L. McCuish, Peter I. Wallace.

Fundy Geological Museum: Kenneth D. Adams, Tim J. Fedak, Katherine M. Goodwin.

Geological Survey of Canada (Atlantic): Jennifer L. Bates, Ross R. Boutilier, Robert A. Fensome, Donald L. Forbes, Lubomir F. Jansa, Christopher D. Jauer, Nelly Koziel, William C. MacMillan, Ronald F. Macnab, R. Andrew MacRae, Peta L. Mudie, Gordon N. Oakey, David J.W. Piper, John Shaw, John W. Shimeld, Rhonda L. Sutherland, Robert B. Taylor, John A. Wade, J.B.W. (Hans) Wielens, Graham L. Williams.

Geological Survey of Canada, Ottawa: Walter R. Roest, Cees R. Van Staal.

Geological Survey of Newfoundland and Labrador: Steven P. Colman-Sadd.

Johnson Geo Centre: Robert G. Grantham (from November 2000).

Kentucky Geological Survey: Stephen Greb.

Mount Allison University: Laing Ferguson.

New Brunswick Department of Natural Resources: Dominique Bérubé, Donald J.J. Carroll, Leslie R. Fyffe, John Langton, Martin E. Marshall, Steven R. McCutcheon, Michael A. Parkhill, Antonius G. Pronk, Arie A. Ruitenberg, Clint St. Peter, Allen A. Seaman, Jacques Thibaut, Reginald A. Wilson.

New Brunswick Museum: Randall F. Miller, Suzanne C. Purdy.

Nova Scotia Department of Natural Resources: Robert C. Boehner, Fred Bonner, John H. Calder, Howard V. Donohoe Jr., Daniel J. Kontak, Garth A. Prime, Ralph R. Stea.

Nova Scotia Museum of Natural History: Robert G. Grantham (to October 2000), Ronald Merrick, Alex A. Wilson.

Oxford University: Connall Mac Niocaill.

Saint Mary's University: Keith Vaughan.

St. Francis Xavier University: J. Brendan Murphy.

University of Alberta: John W.F. Waldron.

University of New Brunswick (Fredericton): Ronald K. Pickerill.

University of North Carolina: Patricia G. Gensel, Brent V. Miller.

University of Prince Edward Island: John R. De Grace.

University of Texas: Christopher R. Scotese.

University of Waterloo: Alan V. Morgan.

References

Colman-Sadd, S. and Scott, S.A., 1994. Newfoundland and Labrador: traveller's guide to the geology and guidebook to stops of interest. Newfoundland Section, Geological Association of Canada.

Cowie, J.W., 1974. The Cambrian of Spitzbergen and Scotland. In: Holland, C.H. (editor), Cambrian of the British Isles, Norden and Spitzbergen. In: Lower Palaeozoic Rocks of the World, v.2, p.123-156; Wiley, London, U.K.

Cox, A. and Doell, R.R., 1960. Review of paleomagnetism. Geological Society of America Bulletin, v.71, p.758 (fig.33).

Dawson, Sir J.W., 1868. Acadian Geology. Second Edition. 694 p.; McMillan and Co., London, U.K.

Ferguson, L., 1988. The fossil cliffs of Joggins, 52 p; Nova Scotia Museum, Halifax, Canada.

Howie, R.D., 1985. Carboniferous evaporites in Atlantic Canada. Compte rendu, Neuvième congrès internationale de stratigraphie et de géologie du Carboniferous, Champaign-Urbana, Illinois, U.S.A., May 1979, v.3, p.131-142 (fig.7).

Leidy, J., 1854. On *Bathynathus borealis*, an extinct saurian of the New Red Sandstone of Prince Edward's Island. Journal of the Academy of Sciences of Philadelphia, 2nd Series, v.2(4), p.327-330 (pl.33).

Scotese, C.R. Paleomap Project. Web site http://www.scotese.com

Van Andel, T.H., 1985. New views on an old planet. Cambridge University Press, Cambridge, U.K. (fig.11.6).

Veevers, J.J., 1990. Tectonic-climatic supercycle in the billion-year plate-tectonic eon: Permian Pangean icehouse alternates with Cretaceous dispersed continents greenhouse. Sedimentary Geology, v.68, p.1-16 (fig.6).

Index

D

E

F

Q

R

S

T

U

V

W

X

Y

Z

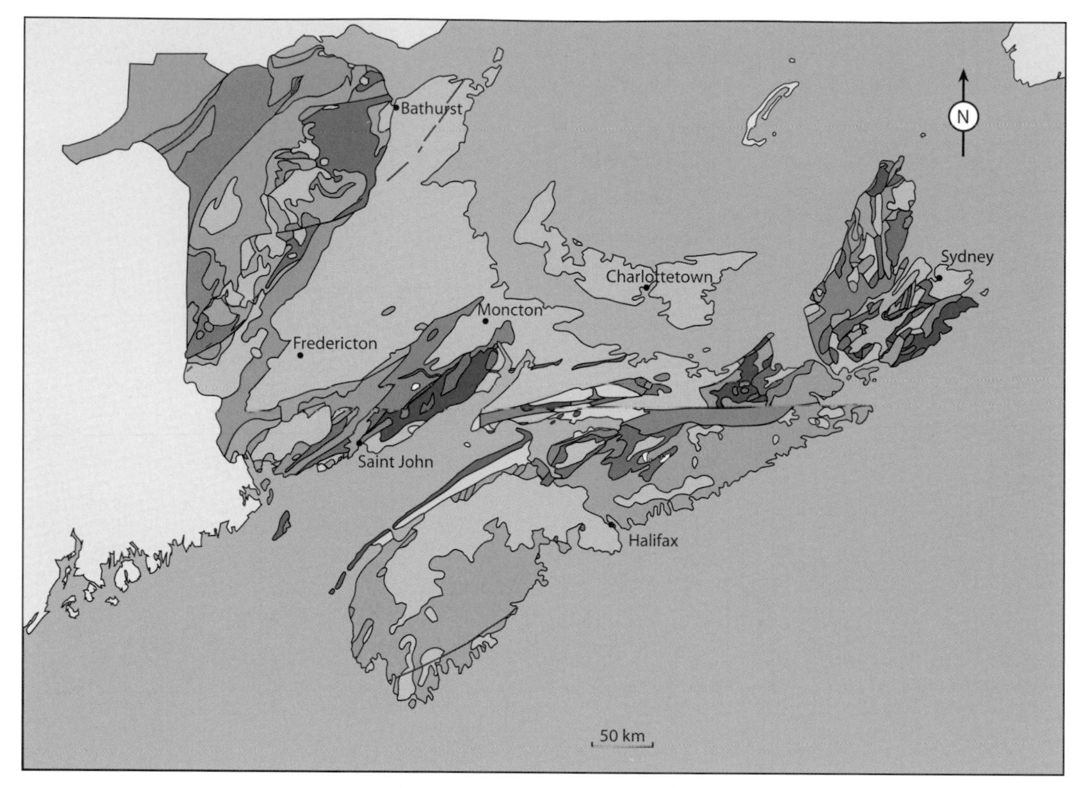

Onshore geological map of the Maritimes.

Mesozoic

- Cretaceous sedimentary rocks
- Early Jurassic dykes
- Early Jurassic volcanic rocks
- Triassic-Jurassic sedimentary rocks

Carboniferous and Permian

- Late Carboniferous-Permian terrestrial sedimentary rocks
- Early Carboniferous marine sedimentary rocks
- Early Carboniferous terrestrial sedimentary rocks
- Early Carboniferous plutonic rocks

Silurian and Devonian

- Silurian and Devonian plutonic rocks
- Devonian volcanic and sedimentary rocks
- Silurian volcanic and sedimentary rocks

Cambrian and Ordovician

- Late Ordovician sedimentary rocks
- Cambrian-Ordovician plutonic rocks
- Ordovician Tetagouche volcanic rocks
- Ordovician Popelogan volcanic rocks
- Cambrian-Ordovician Miramichi sedimentary rocks
- Cambrian-Ordovician Avalon sedimentary and volcanic rocks
- Cambrian-Ordovician Meguma sedimentary rocks

Precambrian

- Precambrian plutonic rocks
- Precambrian Bras d'Or volcanic and sedimentary rocks
- Precambrian Avalonian volcanic and sedimentary rocks
- Precambrian Grenville metamorphic and plutonic rocks

(The colours used in this map are not necessarily the same as those used for equivalent units in the geological maps of chapters 4-7.)

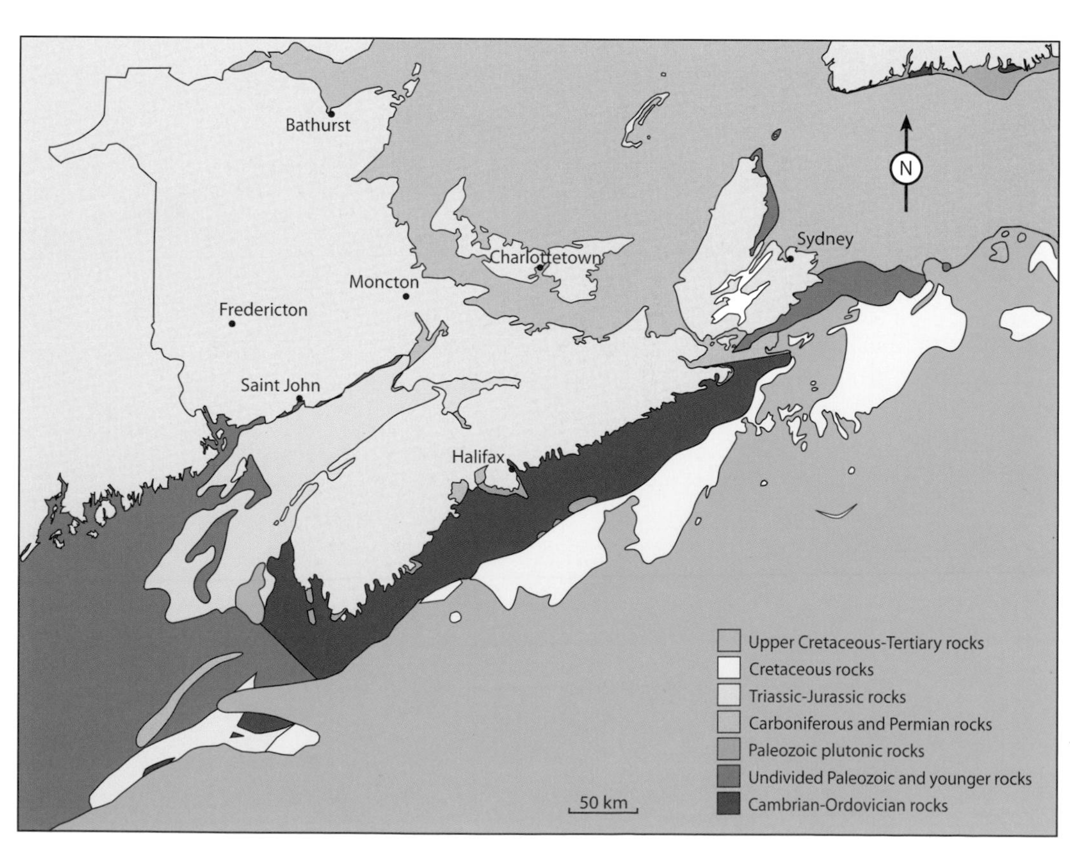

Offshore geological map of the Maritimes.